Hans-Walter Bandemer

Mathematics of Uncertainty

Studies in Fuzziness and Soft Computing, Volume 189

Editor-in-chief
Prof. Janusz Kacprzyk
Systems Research Institute
Polish Academy of Sciences
ul. Newelska 6
01-447 Warsaw
Poland
E-mail: kacprzyk@ibspan.waw.pl

Further volumes of this series
can be found on our homepage:
springeronline.com

Vol. 174. Mircea Negoita, Daniel Neagu,
Vasile Palade
*Computational Intelligence: Engineering of
Hybrid Systems*, 2005
ISBN 3-540-23219-2

Vol. 175. Anna Maria Gil-Lafuente
Fuzzy Logic in Financial Analysis, 2005
ISBN 3-540-23213-3

Vol. 176. Udo Seiffert, Lakhmi C. Jain,
Patric Schweizer (Eds.)
*Bioinformatics Using Computational
Intelligence Paradigms*, 2005
ISBN 3-540-22901-9

Vol. 177. Lipo Wang (Ed.)
*Support Vector Machines: Theory and
Applications*, 2005
ISBN 3-540-24388-7

Vol. 178. Claude Ghaoui, Mitu Jain,
Vivek Bannore, Lakhmi C. Jain (Eds.)
Knowledge-Based Virtual Education, 2005
ISBN 3-540-25045-X

Vol. 179. Mircea Negoita,
Bernd Reusch (Eds.)
*Real World Applications of Computational
Intelligence*, 2005
ISBN 3-540-25006-9

Vol. 180. Wesley Chu,
Tsau Young Lin (Eds.)
Foundations and Advances in Data Mining,
2005
ISBN 3-540-25057-3

Vol. 181. Nadia Nedjah,
Luiza de Macedo Mourelle
Fuzzy Systems Engineering, 2005
ISBN 3-540-25322-X

Vol. 182. John N. Mordeson,
Kiran R. Bhutani, Azriel Rosenfeld
Fuzzy Group Theory, 2005
ISBN 3-540-25072-7

Vol. 183. Larry Bull, Tim Kovacs (Eds.)
Foundations of Learning Classifier Systems,
2005
ISBN 3-540-25073-5

Vol. 184. Barry G. Silverman, Ashlesha Jain,
Ajita Ichalkaranje, Lakhmi C. Jain (Eds.)
*Intelligent Paradigms for Healthcare
Enterprises*, 2005
ISBN 3-540-22903-5

Vol. 185. Spiros Sirmakessis (Ed.)
Knowledge Mining, 2005
ISBN 3-540-25070-0

Vol. 186. Radim Bělohlávek, Vilém
Vychodil
Fuzzy Equational Logic, 2005
ISBN 3-540-26254-7

Vol. 187. Zhong Li, Wolfgang A. Halang,
Guanrong Chen (Eds.)
*Integration of Fuzzy Logic and Chaos
Theory*, 2005
ISBN 3-540-26899-5

Vol. 188. James J. Buckley, Leonard J.
Jowers
Simulating Continuous Fuzzy Systems, 2006
ISBN 3-540-28455-9

Vol. 189. Hans-Walter Bandemer
Mathematics of Uncertainty, 2006
ISBN 3-540-28457-5

Hans-Walter Bandemer

Mathematics of Uncertainty

Ideas, Methods, Application Problems

 Springer

Professor Dr. Hans-Walter Bandemer
Rooseveltstr. 9
06116 Halle
Germany

Originally published in the German language by B.G. Teubner Verlag as "Hans Bandemer:
Ratschläge zum mathematischen Umgang mit Ungewißheit." © B.G. Teubner Verlag | GWV
Fachverlage GmbH, Wiesbaden 1997.

Library of Congress Control Number: 2005932218

ISSN print edition: 1434-9922
ISSN electronic edition: 1860-0808
ISBN-10 3-540-28457-5 Springer Berlin Heidelberg New York
ISBN-13 978-3-540-28457-4 Springer Berlin Heidelberg New York

Springer is a part of Springer Science+Business Media
springeronline.com
© Springer-Verlag Berlin Heidelberg 2006
Printed in The Netherlands

Typesetting: by the authors and TechBooks using a Springer LaTeX macro package

Printed on acid-free paper   SPIN: 11321019   89/TechBooks   5 4 3 2 1 0

Preface

The aim of the book did not require any remarkable changes with respect to contents and presentation for a translation of the German version. Only the references to textbooks in German language were omitted, as a rule, in the translation.

Since this preface is followed by the translation of the foreword to the German edition, which contains also remarks on intention and organization of the present book, it may suffice here to restrict oneself to appreciate all the support the author received in preparation and in writing the present translation.

The author has to thank his former colleagues from the Technical University Bergakademie Freiberg, Professor Dietrich Stoyan, particularly for his critical examination of the German version and his valuable remarks and suggestions, Professor Wolfgang Näther, particularly for his hints to recent publications and his constant readiness for discussion and support, finally, Ingenieur of informatics Irmgard Gugel, for her again excellent work of final technical edition and completion of the manuscript. Moreover, the author acknowledges with thanks the encouragement and occasional support by Professor Reinhard Viertl, Wien, and hints to recent publications by Professor Jürgen Pilz, Klagenfurt.

The author is indebted to Springer-Verlag Heidelberg, represented by Dr. Thomas Ditzinger, for having included the title in the list of his famous Publishing House.

Finally, the author thanks again his wife for her permanent encouragement and understanding consideration, which made the completion also of this translation possible.

Halle   *Hans Bandemer*
May 2005

Contents

1 Introduction 1
 1.1 Application of Mathematics 1
 1.2 The Quality of the Mathematical Treatment 2
  1.2.1 The Quality of the Model 2
  1.2.2 The Quality of the Solving Procedure 5
  1.2.3 The Quality of the Data 6
 1.3 On Model Harmony 7
 1.4 On Information Balance 8

2 Mathematical Representation
of Simple Data and Connections 11
 2.1 Some Elementary Procedures of Data Analysis 11
  2.1.1 Data and Their Representation 11
  2.1.2 Simple Procedures of Data Analysis 13
 2.2 Representation of Functional Relationships Basing on Data ... 17
  2.2.1 Relationships and Data 18
  2.2.2 Interpolation 19
 2.3 Local Approximation 21
  2.3.1 Approximation 22
  2.3.2 Other Approximation Principles 28
  2.3.3 Empirical Smoothing 30
 2.4 Global Approximation 31
  2.4.1 Approximating Functional Relationships 32
  2.4.2 Approximation with Locally Variable Setups 33
  2.4.3 Approximation in Differential and Integral Equations .. 37
 2.5 Approximate Optimization of Empirical Functions 37

3 Specification and Use of Observation Fuzziness 39
 3.1 Specification by Intervals 39
  3.1.1 Simple Error Propagation 39
  3.1.2 Basic Ideas of Interval Mathematics 40

3.2   Specification by Fuzzy Sets 43
    3.2.1   The Idea of a Fuzzy Set 44
    3.2.2   Specification of Fuzzy Sets........................ 46
    3.2.3   Operations with Fuzzy Sets 52
    3.2.4   Connections Via Functions and of Fuzzy Numbers 54
    3.2.5   Fuzzy Relations 59

4   Specification and Use of Uncertain Variability 63
4.1   Chance and Probability 63
    4.1.1   Model Ideas for the Notion Chance 63
    4.1.2   Probability 65
    4.1.3   Random Variables and Their Distributions 69
    4.1.4   Asymptotic Statements 76
4.2   Probabilistic Inference 77
    4.2.1   Samples... 78
    4.2.2   Parameter Estimation.............................. 81
    4.2.3   Testing of Hypotheses............................. 85
    4.2.4   Problems with Imprecise Data 87
4.3   Bayesian Theory 89
    4.3.1   Bayesian Inference................................ 90
    4.3.2   Hierarchical Inference and Robustness 92
    4.3.3   Numerical Problems 94

5   Specification of Vagueness
of Statements on Sets.................................... 97
5.1   Fuzzy Measures 97
    5.1.1   The Idea of a Fuzzy Measure 97
    5.1.2   Forms and Properties of Fuzzy Measures 99
5.2   Simple Inference with Fuzzy Measures 101
    5.2.1   Specification of Partial Ignorance 102
    5.2.2   Possibilistic Inference 105
5.3   Probability and Fuzziness 107
    5.3.1   Probability of Fuzzy Events........................ 108
    5.3.2   Random Fuzzy Sets................................ 111

6   Methods from Qualitative Data Analysis 113
6.1   Crisp Classification of Crisp Data 113
    6.1.1   The Problem of Cluster Analysis 113
    6.1.2   Mathematical Formulation of the Problem 115
    6.1.3   Some Procedures for Crisp Cluster Partition 117
    6.1.4   Clustering with a Mathematical-Statistical Background 120
    6.1.5   Basic Ideas of Neural Networks 122
6.2   Fuzzy Classification of Crisp Data 129
    6.2.1   Fuzzy Cluster 129
    6.2.2   Procedures of Fuzzy Cluster Analysis 130

6.3 Fuzzy Classification of Fuzzy Data 133
    6.3.1 Fuzzy Similarity of Fuzzy Data...................... 133
    6.3.2 The Use of The Concept for Classification 137

7 Evaluation of Functional Relationships 141
7.1 Statistical Regression Analysis 142
    7.1.1 Model Assumptions with Random Dependent Variables 142
    7.1.2 The Problem of Estimation 144
    7.1.3 Discussion of the Model Assumptions 147
    7.1.4 Further A-Priori Knowledge and Assumptions......... 151
    7.1.5 Random Influence in All Variables................... 153
    7.1.6 Local Regression in a Random Field 154
7.2 Fuzzy Evaluation of Functional Relationships 159
    7.2.1 Crisp Data Analysis as a Starting Point 160
    7.2.2 Explorative Evaluation of Functional Relationships 162
    7.2.3 Evaluation with Additional Assumptions 169
    7.2.4 Inference with Fuzzy Parameter Values............... 170

8 Outlook and Conclusions 173

References 177

Index 183

Foreword to the German Edition

Ὁ βίος βραχὺς, ἡ δὲ τέχνη μακρὴ, ὁ δὲ καιρὸς ὀξὺς, ἡ δὲ πεῖρα σφαλερὴ, ἑ δὲ κρίσις χαλεπή. Δεῖ δὲ οὐ μόνον ἑωυτὸν παρέχειν τὰ δέοντα ποιεῦντα, ἀλλὰ καὶ τὸν νοσέοντα, καὶ τοὺς παρεόντας, καὶ τὰ ἔξωθεν.

ΑΦΟΡΙΣΜΟΙ

The present book is an outcome of decades of dealings with users of mathematics and their problems. With the quotation[1] above, adapted to the mathematician, the quintessence of cognition of these dealings can be described. For realizing this it will be helpful to look back some hundred years.

In former times modelling of a practical problem, computing of numerical results for its solution, and expertly interpretation of them were the work of one and the same person: An engineer formulated his mathematical task from his problem, e.g. with simple mathematical terms for a computation by his slide rule or – in more complicated cases – e.g. as a mixed boundary value problem for an approximate solution by series. In every case he was aware that his mathematical model represents a (useful) approximation and his material parameters and measurement values make sense only within a few digits of precision. A confrontation of the results computed by him with the practical problem could be carried out by himself each time immediately.

With growing complexity of the problems tackled the request for "recipes" was expressed more and more perceptibly and gradually satisfied by industrious mathematicians. Because, however, application-oriented mathematics fragmented more and more into subfields, mathematicians as writers of

---

[1] Life is short, art is long, opportunity fugitive, experimenting dangerous, reasoning difficult: it is necessary not only to do oneself what is right, but also to be seconded by the patient, by those who attend him, by external circumstances. Hippokrates, Algorithm I1.
Translation according to The Encyclopaedia Britannica, 14th edition, London – New York 1929, Vol 2, page 97.

"recipe-collections" favoured each time only *one* modelling approach and neglected however other approaches more or less. Moreover, those recipes came off more and more from practical problems and treated only prepared model-setups.

There is, by the way, an analogous development in real cookery books: Whereas hundred years ago the young housewife was still said how a hen to be bought for a certain dish should look like (with respect to age, weight, and constitution) and how it is to be treated before cooking, today one reads laconically: take a hen.

By the explosively increasing efficiency of electronical computers for general scientific requirements and by the deluge of available software packages the mathematical recipe system was increased extremely. As is well known, it is the pride of any software seller to have collected in his package, possibly, all known appropriate procedures. With this tendency to perfection the programmer cannot be presupposed any longer to have an overview over the concrete application situations. The necessary use of software by the applying engineer *separates* him necessarily from his practical problem. (The engineer stands here only as a representative and an example for a scientifically working user.) The comfortability of the software tempts him additionally to replace his expertly reflection on a model by an easy-made choice from a catalogue of offered models, which should be useful for the problem only in few cases. By the psycologically designed output of the results as colourful diagrams the user has, as a rule, hardly a chance for a critical assessment of the results with respect to relevance, reliability, and precision.

At this point the matter of the present book starts. The reader is assumed to have access to hard- and software suitable for his problems. The book is so neither a handbook nor a textbook for the application of mathematics in special practical situations. Instead of, it should represent a recommendation for the user, in which he is said what he should really take into account in application of mathematics together with this modern equipment in order to come to a reasonable treatment of his practical problem. This topic includes the choice of such a *mathematical model* that is adequate to his practical situation and to the requirements of his aim with respect to complexity and mathematical tractability, as well as the choice of an *effective solving procedure* for the mathematical task, and, finally, the choice of such criteria, which allow to *assess realistically* the reliability of his starting values and the dependability and precision of the results.

When so, in the following, procedures are presented explicitly, then only in principle or as typical simple examples. There is no claim made with respect to any completeness. Because, even though the procedures are permanently improved and increased, the problems *with their application* remain, in principle, the same.

Already rather early in application of mathematics a certain separation of competence took place. Interest in concrete mathematical modelling exists essentially within the *applying* discipline. Mathematicians, however, are concentrated on the development of mathematical solving procedures and, in fact, each only in his respective subfield as analysis, algebra, stochastics, or numerics.

Because each of these subfields has developed its own traditional denotation, it was not trivial to find a workable compromise for the presentation in the present book, simultaneously avoiding a set of different types of characters being too extensive.

Therefore, as long as possible, no special type was used for common sets, variables, and functions. If with *vectors* and *matrices* their character should be emphasized, e.g. because their connection rules are used, then they are denoted by bold-faced letters, e.g. $\mathbf{x}$, $\mathbf{B}$. In this context the *components* or *elements* can be also random variables or fuzzy sets. Common n-tuples, e.g. multidimensional parameters, are not stressed as vectors.

Random variables are denoted by sanserif-letters, e.g. G, T, also if they have the character of vectors or matrices, but this property do play only a secondary role in the given context. For random errors, however, the traditional exception (ϵ) is used to guarantee the connection to the usual textbook literature.

For *fuzzy sets*, unknown yet in many places, letters in script ($\mathcal{A}, \mathcal{B}$) are used, if not their character as vectors or matrices should be emphasized.

Finally, occasionally for special sets and systems of sets also shadow letters are introduced, as $I\!R$ for the Euclidean space, $I\!P$ for the power set of a set, and $I\!N$ for the set of natural numbers.

The formulae used should either remind of facts learned from textbooks or specify uniquely newly presented mathematical ideas. In every case the most important details are explained within the text. Hence, if someone does not understand the formulae immediately, then he may pass over them in his first reading without any loss of comprehension for the facts following and concentrate on the text. Surely, he will then, in repeated reading, feel that it means a thing to him.

Cross-references occur in usual manner to chapters, sections, and numbers of formulae.

If it is referred to common textbooks and in this context also some titles are named, then this does not mean that only these are recommended. The author did not take the trouble to look through the each time immense set of literature; the cited books lay on his desk and he has used them to look up.

The author has, in his academic life, worked through different fields of mathematics and in this course gathered varied experiences in application. In the present book he wants to convey of these to a wide section of users, but without to emphasize natural philosophical insights. A standard mathematical course for students of engineering or of another applying discipline at the graduate level should suffice to understand the presentation, certain experiences with application of mathematics are desirable.

Especially such fields of mathematics are dealt with, in which uncertainty, variability, and imprecision of data are of importance, e.g. if they are to be obtained from practical situations.

Starting with values of functions, from observations or measurements, at first problems of *interpolation* and *approximation* are treated. Then *observation fuzziness* is considered and mathematically specified, which can influence essentially preciseness and reliability of statements on functional relationships. Then the notions of randomness and probability are examined as a model for the *variability* of observation and measurement results and their use is considered critically. Finally, also a mathematical approach is presented for the *vagueness* of data, and observation fuzziness and variability are handled under a mutual point of view.

After this supply of basing ideas the book turns to some methods of qualitative data analysis (cluster analysis and classification) and of evaluation of functional relationships (regression analysis and quantitative fuzzy data analysis).

If a mathematical approach, followed up by a reader for years, remains unmentioned or occurs only by a hint in the present book, then the author asks this reader to be lenient, because for the matter of the book a selection of topics must be sufficient.

Hints to errors or shortcomings (unavoidable even with great care) are always welcome.

The book was written essentially unnoticed by colleagues and without any support by a scientific staff. Hence the author has to thank only the coordinator of his graduate college *Spatial statistics*, Dr. Martina Lorenz, and his last graduate student for a doctorate , Dr. Silvio Hartmann, each for a critical examination of the manuscript, and Dr. Wolfgang Fleischer for his kind allowance to reprint a figure from his book (see Fig. 2.1 of the present book). The technical final edition and completion was again, as with former manuscripts of the author, in the reliable hands of Irmgard Gugel. The author thanks her cordially for her excellent work.

The author is indebted to Teubner-Verlag, especially to Jürgen Weiß, for his insistence, with which he has pressed him for years to write the manuscript, and for his understanding cooperation in the genesis of the book.

Finally the author thanks his wife for her permanent encouragement and understanding consideration, which made the completion of the book possible.

Halle   *Hans Bandemer*
April 1997

1

Introduction

1.1 Application of Mathematics

The central problem of every science, as is well known, is the *determination* of conclusions from realized conditions, i.e. producing statements of the form

IF ... THEN ...

According to the degree of formalization of the branch of science in question these statements are presented as more or less precise verbal concentrations of experiences up to mathematically formulated laws of nature, e.g. the law of falling bodies.

In every case mathematics serve for ordering of thoughts and for ,,ready-making" of reasoning. Hence mathematics provides *brain tools, by which something can be managed intellectually*, likewise the hand tools, which a manual worker uses to make his work easier or even anyhow possible. Thus the mathematician corresponds to the tool maker in some respects.

Application of brain tools to real problems has some similarity with the use of hand tools:

1. You must get clear, what you are really want to aim at, e.g. you must fix the *aim* (and the purpose) you want to reach by your investigation using mathematical means.

2. Next you have to fix the *quality requirements*, which the result of your investigation has to fulfill. This becomes clear if you compare the requirements on a box for rejects with those on a cabinet. This is the same case with respect to the results of a mathematical problem. Fixing of quality is important, it determines the tools and the requirements necessary for a satisfying solution of the problem.

Now you can start with an appropriate mathematical treatment of the given problem. First you have to translate the structure of the real problem and the realized conditions into the abstract language of mathematics.

This translation divides into three components:
a) the choice of a *mathematical model;*

b) the choice of a mathematical procedure for the conclusions; shortly a *mathematical solving procedure;*
c) the description of the given situation by *mathematically specified data.*

For the "quality" of the mathematical treatment and of its mathematical results there is a *rule of the weakest chain-link:*
The result in every given case is, with respect to its meaningfulness (e.g. relevance, reliability, accuracy, precision), at most as good as the weakest "quality" of the three components.

The notion "quality" points in the same direction for each of the three components, but it must be defined in different ways for each of them, and, unfortunately, it can not, in general, be recorded quantitatively. The next section will be concerned with this notion in more detail.

1.2 The Quality of the Mathematical Treatment

1.2.1 The Quality of the Model

In modelling the notion *quality* will be interpreted as *adequacy.* Modelling of a real problem is not unique, there are, as a rule, many possibilities to formulate mathematically the structure of the given problem. One can interprete *adequacy* as *sufficient* correspondence with reality. In this statement sufficiency depends on the aim and the purpose of the investigation. (Hence fixing of aim and purpose was put on the first place above.)

As a simple example consider the playing at dice.

To predict the result of a throw *in the long run* of a game a simple stochastic model will suffice, by which you can decide, if necessary by a test, *whether the used dice is unfair.*

To predict the result of one single throw, considered as *a physical phenomenon* one needs a large expenditure of measurement instruments to record all the influencing factors of the dice, with respect to the throwing technique (e.g. starting acceleration and throwing direction) and impact surface (e.g. elasticity and roughness). Possibly in this case a system of partial differential equations with many variables may suffice for an exact prediction, but there remain some doubts, whether this may do.

Besides on orientation on aim and purpose of the investigation the choice of the model depends also on the different complexity and mathematical and numerical manageability as well as on the practical transparency and interpretability of the models to choose from. Frequently the choice of the model is also influenced by the order of magnitude, in which the problem is to be handled. All these aspects will be explained by examples in respective situations later on.

The choice of the model is a *creative* main problem of the user himself; he can be supported by the mathematician at most by providing of brain tools. In this way in a given context, e.g., a model from the field of differential equations becomes possible as well as one from regression theory.

An assessment of model quality, of *adequacy*, on theoretical grounds by means of mathematics, is *in principle impossible*; nevertheless it is tried again and again, e.g., by means of statistical test theory. In doing so the user forgets that in this manner he decides only, whether the given data are sufficiently probable, if some special assumptions are valid, whereas a set of other assumptions is taken for granted.

A proper assessment is only possible by confrontation of the results of the theoretical investigation with practical findings. But even in the case of sufficient agreement the conclusion that the used model is adequate do not need to be valid. In many cases it suffices to state the *usefulness* of the model. What is meant by these remarks in practice will be explained in the following example.

As is well known, in the flat rolling process the material, e.g. steel, is pushed through the opening between two rolls in order to reduce the thickness of the material. Usually experts (e.g. HENSEL/SPITTEL (1978)) assume that this process can be modelled as a process of continuous compressing: As soon as a small part of the material reaches the opening it is pressed by the rolls from above and from below, i.e. the material is compressed. Hence the mathematical models for flat rolling are taken from those for compressing. In the fifties of the last century an applied mathematician had the idea to consider the treated material under the high pressure of the rolls as a viscous fluid and to model the rolling process by hydrodynamics (see e.g. KNESCHKE/BANDEMER (1964)).

Denote $p(x)$ the pressure distribution and $u(x,y)$ the speed component of the material particles parallel to the rolling plan, then the movement within the opening of the rolls can be modelled by means of the hydrodynamic theory of fluid friction using the differential equation

$$\frac{\partial^2 u}{\partial y^2} = \frac{1}{\eta}\frac{dp}{dx} \; , \qquad (1.1)$$

where η means the dynamical viscosity of the material to be rolled, which stands in the opening under a certain fluid stress. The used differential equation was obtained by extensive simplification (neglecting unimportant terms) from the NAVIER-STOKES differential equations.

To investigate the adequacy of this model the technical fact was used that the steel within the mould solidifies not as a whole but in layers following the temperature gradient. This can leave a linear structure within the ingot to be rolled: a net of parallel planes (Seigerebenen). If, for that investigation of adequacy, the rolling direction is chosen as being orthogonal to these planes, then the distortion of the cross-section plane of the rolled material can be made visible experimentally (by cutting the rolled material orthogonally to the cross-section plane and corroding the obtained sectional surface plane). At the same time the distortion of the cross-section plane was computed, assuming that the model with the differential equation (1.1) is valid. This was

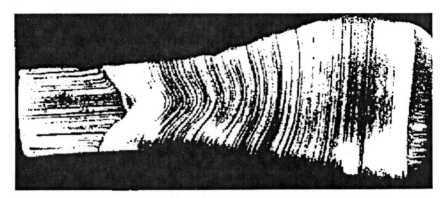

Fig. 1.1. Move lines in a specimen (By courtesy of the Institut für Bildsame Formgebung der RWTH Aachen)

effected by solving the corresponding integral equation of the first kind numerically using an integraph, as it is called, a mechanical tool for integration of a given function. From the solution a figure was drawn showing the momentary position of the material particles on the cross-section surface plane after different time lags.

Figure 1.1 shows the move lines from the input to the output of the rolling train (from the right to the left). It is taken from a paper by Kneschke (KNESCHKE (1967)).

Figure 1.2. shows the result of the computation for the model with assumed parameters and is taken from the paper KNESCHKE/BANDEMER (1964). The pictures correspond with each other with respect to the mathematical form of

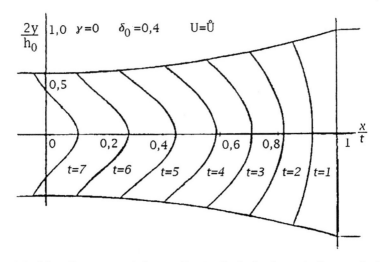

Fig. 1.2. Move lines computed according to the hydrodynamic theory of rolling

the curves amazingly well. If the assumption of continuous compression would be valid, the straight move line of the input must be followed by straight move lines throughout and after rolling. Nevertheless the hydrodynamic rolling theory was not accepted by the engineers, because they did not see any possibility to measure the dynamic viscosity η, which depends not only on the material but also on temperature and pressure.

1.2.2 The Quality of the Solving Procedure

The *quality* of a solving procedure may be considered as its property to provide the solution of the given problem with *sufficient precision* and *tenable* (appropriate) *expenditure.*

The *assessment* of this quality can be effected after interpretation of the corresponding result. *Iteration procedures,* e.g., will be assessed with respect to their *numerical stability,* i.e., if such a procedure had come, in a certain step by a more or less gross numerical mistake, to a weaker approximate solution, then the procedure should approach, in the next steps, the exact solution again. Besides, the procedure can be assessed by its *speed of convergence,* i.e., *how fast* it approaches a sufficient solution, if no gross mistakes are made in computing. Finally, *bounds for the solution* are of interest, when the procedure was stopped at some step, e.g., when there was no more change in the approximate solutions.

In mathematical statistics the estimation procedures, used in this field, are assessed, among others, according to *unbiasedness* and *variance of estimation,* i.e., according to their properties *in frequent use.*

In every case the assessment follows criteria of precision and efforts. The investigation, the comparison and hence the construction of such solving procedures are essential problems of mathematicians.

The user should, in every case, be interested in corresponding statements with respect to these properties for procedures suggested to him, he should look for them in the literature or demand them from the supplier of the procedures.

All these properties of the procedures are stated *on certain conditions* with respect to the model and the data used. These conditions are fulfilled in practical cases at most approximately. Hence for the user *statements of robustness,* as they are called, are of importance. These statements explain, how strongly, or better how weakly, deviations from the ideal conditions have an effect on the properties of the procedures.

In mathematical statistics (see Sect. 4.2) the arithmetic mean of a sample is suggested as a *good* estimation of the expectation value EX of a random variable X. This estimation is of a good quality, its variance is minimal. However, if gross mismeasurements (a common form of outliers) are expected, it may be advantageous to *ignore* such measurements in averaging, which lie very far from the arithmetic mean. The estimation will be (presumably insignificantly) worse in quality (the variance grows a bit), however possibly occuring outliers can no longer bias the estimation, at least not so heavily. So the estimation is made *robust* against the occurence of (some few) outliers. This strategy is an example for the construction of *robust* statistics, as they are called. A further frequently used strategy for robustification consists in a variable weighting of the sample elements in averaging (see for this problem area e.g. HUBER (1981)).

1.2.3 The Quality of the Data

The quality of the data is assessed by how *reliably* they characterize the given problem. So the question is, whether they are the *right* data: Are they suitable to reflect the given problem in the model or are they only minor matters of neglectable importance? Do they belong really to the given practical problem or do they describe a quite different situation? Are they carefully and responsibly obtained and *specified*, suitably for the problem, as numbers or as other mathematical objects?

According to the rule of the weakest chain-link the answers of the questions decide essentially on the quality of the whole investigation.

Seldom the common user is conscious of how often gross errors (e.g. outliers), imaginative data or manipulated values occur in his data. The application of methods from data analysis (see Sect. 2.1) for detection and possible deletion of such suspicious data is an only weak compromise measure. This is, as a rule, inferior to a technical and critical examination of the data.

Moreover, already in specifying the data, mostly unconsciously, some "preciseness" is introduced. So, by the number of given digits it is insinuated implicitly that the last given digit is precise to one half of its unit. When considering what the datum is to express this statement is seldom valid.

Reading a room thermometer results in the temperature in the room at most in the surroundings of the thermometer with a reading accuracy of half a degree. The spatial variations of temperature within the whole room, however, should be of a wider range.

This wrong assessment of data precision should be of more serious consequences when the data are handed over to a computer, which brings them up to its computational accuracy by adding zeros for the lacking digits. Then, however, the results of the computation are only valid for those "highly precised" data!

Since the quality of the data is scarcely considered in present-day applications of mathematics and does not play any role in common textbooks and software tools, in the present book the *main attention* is devoted to this chain-link. Here some reserves lie for a realistic assessment of the computed results, on which factual decisions are to be based, as well as as a source, from which one can scoop, if an adequate model is not (yet) found for the given facts. In the following chapters the different principal practical initial positions are considered in the interaction of model, procedure and data.

The main scope will be occupied by the investigation of such relations, for which mathematics plays a central role. The text will start with a purely empirical representation without any use of mathematical formulae. Then the consideration will pass over to specify the problems by such formulae, via simple approximation to representations with different backgrounds, as in mathematical statistics and fuzzy theory. In parallel mathematical means are described to specify the uncertainty of data and hence of the results for practical facts. The mathematical procedures are treated only in principle or as examples. It is assumed that the reader has access to efficient hardware and software, hence elaborated numerical examples can be omitted.

1.3 On Model Harmony

The choice of a model implies always also the fixing of the interpretation of the data used, the choice of an appropriate type of solving procedure and, finally, also of the form of statement, in which the solution is to be presented.

In mathematics the following *model backgrounds* are available:

numerical approximation theory;
interval mathematics;
probability theory and mathematical statistics;
and, in these days also, fuzzy set theory.

Thereby the user comes into the position to choose such a model background, which seems to him the most appropriate, depending on the given problem, his general background knowledge and his decision on adequacy.

In doing so the different forms of solution presentation get transparent, following the principle of *model harmony:*

approximate data as the starting point supply *approximate* statements;
interval data as the starting point supply statements in the form of *intervals*;
fuzzy data as the starting point supply statements in form of *fuzzy sets*;
probabilistic assumptions on the origin of the data supply statements with *probabilistic character*.

As an example for the case last mentioned the situation is considered that the given data are interpreted as realizations of random variables, on the distribution of which certain assumptions are put. The results of the solving procedure are then interpreted as probability statements, e.g. as confidence regions for some unknown parameters.

Frequently, however, in applications, even sometimes in contributions to serious scientific journals, the reader is confronted with an eclectic conduct of the authors. In those contributions approaches are *mixed* ignoring that the character of the conclusions and statements depend on the *assumed* character of the data used. In this way, e.g., procedures from different backgrounds are applied to the same data and the results are compared *numerically* taking no notice of what type of data and with which background these procedures are developed. Investigations of such a kind are in the best case useless and in the worst case misleading.

A *combination* of approaches, however, is sometimes possible. In Subsect. 7.2.4 the case is presented that the data are realizations of *random variables* and, simultaneously, specified as *fuzzy sets*. There are already special procedures available to handle this combination (see e.g. VIERTL (1996)).

Hence, in the following chapters always the background of the model and the type of the mathematically formulated problem are explained, for which the given model and the solving procedure make sense.

1.4 On Information Balance

The mathematical treatment of a practical problem can be taken as "information processing" in a wider sense of the wording:
Information on the problem from different sources is to be "assembled" or brought together in such a way that the problem question can be answered "sufficiently precise".

As sources of information are mainly possible:

1. *Sure knowledge* from practically confirmed theories and reliable experiences as is available in the literature and with experts. Hence it is always worth to read and to ask, before tackling a problem.
2. Knowledge from the neighbourhood of the given problem on some connections, conditions, previous experiences, which can be condensed to reasonable assumptions.
3. Finally empirically and experimentally obtained results, which will be called (a special form of) *data* later on.

A "Golden rule of problem solving" can be phrased:
Information necessary for the problem solution and not (yet) known or not included, must be obtained requiring expenditure of time, material, and manpower.

Hence it is always useful to draw up a *balance sheet* for the information:

Information already *available* from the three sources above stands on the *credit side*, whereas information *necessary* for the solution enters the *debit side*. On the balance you see then what is yet required as well as what is not necessary for the given problem. In this balancing the purpose and the aim of the investigation as well as the quality requirements on the solution, as discussed in the preceding subsections, will play a central role.

Naturally, for a balance sheet one needs some "monetary unit" to compare and balance the single entries on the credit and debit sides with each other. This demand raises the problem of a mathematical definition and specification of information. But for the further practical considerations any specification and measuring of a certain quantity called "information" will be *not* of importance and hence not treated. Because this would lead to an academic discussion of a mathematical *notion of information*, e.g. via entropy measures. In certain fields of mathematics there are proposals for this purpose. However, in the present text this notion is looked upon as an allegorical one. Information processing means here: *including* of available information into the given problem by *mathematical* specifications, and *recording* of knowledge on the problem, e.g. by *formulae and rules*, and on the environment of the problem by *mathematical assumptions*, and of the data characterizing the situation by *mathematical objects* (numbers, vectors, functions, etc.). Not until the available information is specified and recorded in such a way one can turn over to question how this information can be *combined* and *used* to solve the given problem. But these considerations are already part of the *design* of the investigation, which has to contain also the chosen solving procedure, if and when the information is specified and recorded.

In the following chapters different situations for the information, for the problem aim, and the quality demands on the solution, together with requirement lists to the data are treated, with which this basic position will be illustrated.

2

Mathematical Representation
of Simple Data and Connections

In this chapter mathematics is used only as a means for *representing* data and their connections. Experts' ideas on possible models and assumptions on the genesis of the data and their uncertainty are omitted. In the chapter some techniques are presented, which will form the basis of solving procedures later on in the book, when they are considered under additional assumptions on the model and the data genesis.

2.1 Some Elementary Procedures of Data Analysis

Before one can speak of data analysis, it must be clear, what is meant by a datum.

2.1.1 Data and Their Representation

Up to now the expression "datum" was used, but its concept was left vague. Within mathematics usually a *datum* is understood as a real number, obtained as a result of some measurement or observation and to be introduced into a mathematical procedure.

For a better inclusion of really practical situations the notion will now be mathematically formulated more generally and exactly.

The wording "datum" means, literally, "something actually given". It makes sense only in a certain context and expresses that "something" was found in a state characterized by just this datum. Obviously, such a datum contains information only if there are at least two different possibilities for the state of the "something" in question. Hence one can consider every datum as a realization of a certain variable in a set of values, called the *universe of discourse*, and reflecting these possibilities for the state in the given context.

The first task in a mathematical modelling of such an affair consists in a mathematical representation of the possible data simultaneously specifying a suitable universe. For illustration some special cases will be mentioned.

In the simplest case the datum only reflects whether a certain characteristic *property* (an attribute, a feature) is present or absent. The suitable universe is then any two-point-set. If, as usual, the set $\{0, 1\}$ is chosen as the universe in this case, one must be clear that the two numbers are no "measure values", but evaluate only the *truth of the statement*:

The something has the certain property.

In this context the number 1 means that the statement is *true* and the number 0 that it is *false*. Hence the variable over this universe is called a *logical variable* (see, e.g. JAMBU (1991)).

If there are several possibilities for a shaping of the property, e.g. several possible places of origin or types of a product, several possible answers to a questionnaire, then the suitable universe will be a *finite set*, the members of which are usually chosen as *characters*. The likewise usual choice of natural numbers is in general rather problematic, because the numbers, in the given context, are only distinguishing marks, denoting *categories* and not implying neither a sequence nor a scale of values. Hence variables with such a "range" are also called *categorial variables*.

An interesting special case arises if the different possibilities are *grades* (degrees, shades, nuances) of a considered property, e.g. the number of children in a family, but also nuances of hair colour (very fair, fair, light brown, etc.) or degrees of satisfaction (not contended, little ..., more or less ..., moderate ..., very ...). When the use of numbers makes sense as a measure for childfriendliness, their use for nuances and grades is absolutely arbitrary and suggests frequently the existence of a "measurable distance" between them. This problem is considered in more detail in Sect. 3.2 Variables of this type are often called *ordinal variables* , in contrast with categorial data, in which such an order is *not existent*, the sequence of the categories may be arbitrary. Categorical data of the latter type are sometimes called *nominal* ones. (See also AGRESTI (1990)).

When considering *observations* or measurements, the result in a concrete situation is usually given by *numbers*. Although every result of an observation or a measurement can be given only with *finite many*, mostly very few digits, the whole *real axis* is chosen as an appropriate universe, which can supply relevant advantages for a theoretical treatment. Usually such variables are called *quantitative variables*.

If the observation or measurement is performed continuously in time or space (e.g. temperature records, spectrograms, geological profiles), then every result is presented by a trajectory or a (hyper-)surface, resp. The corresponding variables are called *processes* and *regionalized variables*, resp. Then the universe can be chosen e.g. as an appropriate space of functions (e.g. a HILBERT-space) ore some finite-dimensional approximation of it (a so-called "setup"). The transition to *time series* (observations at a sequence of times) or to *grids* (observations at a grid in area or in space) leads to presentations by vectors or by matrices, resp.

Here usually the opinion ends what may be understood as a type of data. But with respect to new developments within mathematics already at this point two further types of data will be mentioned.

Also *grey tone pictures* (and colour pictures) may be considered as data, since they reflect each "the state of a certain something", e.g. X-ray records, projections of transparent sections taken from cellular tissues, two-dimensional projections of three-dimensional particles recorded by optical devices, or some scenes scanned by some tv-equipment and to be recognized. A possible representation of a grey tone picture is a basing area (the picture frame) within which a function is defined reflecting the grey tone at each single point of the picture by a numerical value. Such a representation can often be not appropriate, e.g. if the semantic *picture content* must be the starting point for the investigation (e.g. extent, shape and position of a tumour). Whether, in such a case, a partial set of the set of all possible grey tone functions or whether only the pixel field matrices of the finite many distinguishable grey tones should form the universe, depends on the practical problem standing behind the modelling task for the data.

Finally, every *opinion of an expert* or of a panel of experts can be interpreted as a datum, when it is related to the state of a given "something" (e.g. the state of the market, a certain investment, the weather forecast, etc.) These opinions are usually uttered as *statements, rules and conclusions*. The mathematical formulation will then be taken from mathematical logic. The appropriate universe is then chosen as a set of such statements, rules and conclusions, being available as possible "values" for the opinions of the experts. Simple examples are questionnaires with a given range of anwers to each single question. The usual coding of experts' opinions by numbers is a but rather doubtful and arbitrary proceeding. A more problem-oriented mathematical representation of grey tone pictures and experts' opinions will be presented in Sect. 3.2

Moreover, one should notice that every datum is connected with uncertainty of different origin and kind. Examples are given by considering the larger or smaller variations of the state of the something in comparable situations *or* the unsharpness of every in some respect somehow rough measuring tool *or* the differences in the understanding of words and notions among experts or people answering a questionnaire, e.g. with respect to the intended sense of the different answers. Not only in each of these cases the respective state of the something in question cannot be reflected *totally* by any mathematical specification. Disregard of all uncertainty, however, can *essentially influence* the result and the conclusions of any data analysis, this will become clear in Sect. 3.2

2.1.2 Simple Procedures of Data Analysis

Data analysis consists of an investigation and an evaluation of the given data, and of the drawing of conclusions from the data, and of the evaluation of these conclusions. Data analysis is performed in several stages of increasing complexity, of which stages now only the first two are considered for the moment.

In the first stage the data have to undergo first a critical inspection from the standpoint of the branch of science the given problem is taken from. Even if the size of the data material is rather extensive, one should at least require some impression on the *reliability of the data*. By this one has to decide, whether the data *really* describe *comparable* states of the "something". Naturally, it would be best, if the "data analyst" has obtained the data *himself*, or if he had, at least, got a personal idea on the procedure of the "data capture" so that he can assess its care of performance and its level of precision. If he finds, in his inspection, such data that are inexplicable or unexpected from the standpoint of his branch of science, he should again look at the circumstances under which these data were obtained and try to confirm or correct them. In the case of confirmation he may come to some new insight with respect to the reflected state. Mostly, however, he will find that an error or a mistake in the data capture occured. Within statistics such data are usually called *outliers*, because they lie, as a rule, far away from the bulk of the data. It is unwise, however, *without any consideration* to declare all those doubtful data outliers and generally to delete them, or perhaps to let the decision to the software, which offers some formal criteria for this purpose.

Moreover, the data should be inspected with respect to possibilities of effected *manipulation*. This case is more frequent than people believe. The person, who has to obtain the data, or the data analyst or the client of the project possibly may have some personal interest that the data lie in a certain field, e.g. because this is connected with the quality of his work or with the profit of his enterprise, respective, or the person is so apathetic towards the data that he avoids the effort of data capture, and he records data according to his imagination seeming plausible to him. Obviously, in both cases the data are worthless for a scientific investigation.

A further reflection on the data concerns their *essentiality, credibility* and *trustworthiness*, e.g. whether the data reflect the state of the "something" *truly*, although they are recorded reliably. Do the data characterize the state *accurately*, or, e.g., only *irrelevant* symptoms were recorded? Also with respect to the substance of experts' opinions this fact is important and suggests the introduction of competency weights and truth-values for their statements.

Not until these inspections the *second* stage of data analysis with mathematical means will make sense. The procedures of simple (mathematical) analysis consists in *arrangements* (sorting and plotting) and suitable *transformations* of the data (see, e.g., POLASEK (1994)).

The very first arrangement of the data is performed, as a rule, with respect to *frequency*. A single but important property of the "something" is considered and evaluated at a time. A *frequency analysis* deals with "usual" states of the "something", e.g. with respect to its expected states in future. For this purpose "similar" data are combined in *classes*, the frequency of which is determined. If a stochastic background is supposed for the genesis of the data, the frequency distribution is taken as a hint to a possibly existing probability distribution. Data far away from the bulk of the data will be handled as potential outliers.

The main aim of data analysis, however, is the *search for pattern* or structures in the data , from which hints may be obtained about possible mathematical models, which may be used for further conclusions.

In *this* chapter procedures are presented, for which it is assumed that all data introduced are *reliable* and *trustworthy*, and that these data are represented by real numbers *exactly* or, in the multidimensional case, by real-valued vectors likewise. This means, among others, the assumption that there is not any information available with respect to the possible uncertainty or inaccuracy of the data, at least that such information is not used. In later chapters, when such information will play its role, data of this kind will be called *pseudo-exact* data.

Since the following procedures do not use any information with respect to possible models, to the background and to possible side informations and prior knowledge on the given practical situation they require a relatively *large* size of the given data, according to the Golden rule in Sect. 1.4.

The main strategy will be a "visualization" of the data: two- or three-dimensional projections of the data vectors are made visible as points on the screen and are inspected by eye with respect to *patterns* or *structures*.

With respect to pattern two large groups are distinguished:

So a "data cloud" can break into "partial clouds" more or less clearly. These partial clouds are called *clusters* , and one looks for scientific explanations why and how they occur. In this way a doctor comes to special diagnoses, by which diseases, the symptoms of which may be rather similar, can be distinguished in future cases.

In the other case the data points may be grouped around mathematical structures as *functions* or *surfaces*, which could give a hint that such a regularity can be used as a mathematical model.

Frequently those structures, clusters or functions are not visible at once, but only when the data have been transformed first. Such *transformations* are sometimes suggested by the context, usually the software for data analysis offers a wide variety for this purpose. For this approach the assumption is essential that the data are specified *exactly*, because the transformations can distort any small deviation of the original in a non-transparent manner. To give an impression of such distortions a nice example is reproduced here taken from the book NAGEL/WERNECKE/FLEISCHER (1994).

If one accepts a transformation offer from the software and finds so structures in the data material, then one should really consider the obtained results whether they make sense and can be interpreted in the given situation. It is quite possible that there are several mathematical representations for the structure in question, among which it oscillates, when more data are obtained. Moreover, one remember the advice for a possibly *simple* mathematical representation. A further possibility to realize structures consists in

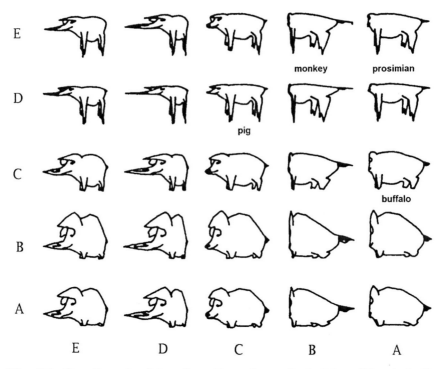

Fig. 2.1. Two-dimensional transformations of an animal picture. The pig in the centre of the figure is the original picture. The characters mark the type of transformation of the corresponding axes: A = square, B = exponential, C = linear, D = logarithmic, E = square root. (By courtesy of the authors of the book cited above, see page 183 of the book)

the construction of *contingency tables*, as they are called. In the simplest case two different properties, say A and B, of the objects to be investigated (states of the "something") are considered and the elements of the corresponding universes are collected in classes K_i and L_j, respectively, of neighbouring or similar values. (Naturally, with nominal data the categories form the classes.) The frequencies m_{ij} found for the combinations (K_i, L_j) will be inserted into the boxes of the table. These frequencies can be interpreted, using a stochastic background, as estimates of the corresponding probabilities, or they will give only a hint to a possibly existing connection, combining somehow the classes with the highest frequencies, between the two properties, which should be investigated further (see, e.g., AGRESTI (1990)).

If the data vector (of the properties, which should be considered simultaneously) has a high dimension, then the confrontations in pairs or in threes remain possible to a certain extent (sometimes offered by a field of sixteen on the screen), but any track is increasingly lost with the number of components of the data vector.

Table 2.1. Example of a contingency table. m_{1+} means the sum of the m_{1j}, m_{+1} the sum of the m_{i1}, etc., and m_{++} the total sum for all cases

$B \backslash A$	K_1	K_2	K_3	K_4	$\sum$
L_1	m_{11}	m_{12}	m_{13}	m_{14}	m_{1+}
L_2	m_{21}	m_{22}	m_{23}	m_{24}	m_{2+}
L_3	m_{31}	m_{32}	m_{33}	m_{34}	m_{3+}
$\sum$	m_{+1}	m_{+2}	m_{+3}	m_{+4}	m_{++}

For this case a projection technique was developed (see, e.g., FRIEDMAN/ STUETZLE (1981) and HUBER (1985)). The high-dimensional real-valued data are projected, from different standpoints in succession, into the plane or the three-dimensional space (on the screen). The aim of this procedure is to observe a structure when reaching a certain angle of vision, e.g. some disintegration of the point cloud or its grouping around a mathematical dependence representation, a curve or a surface. One may imagine the technique like as on a flight around the Galaxy, on which flight the lentil shape of it becomes clear especially under a certain angle. This technique was perfected in a manner that by a system, similar to that used in orbit runs of spacecrafts near to the earth, at each time shifted by a small angle, an animation of the cloud of the projected data is effected on the screen. The observing data analyst can stop the picture at any time and register the present direction of the projection, by which proceeding the structure found can be mathematically recorded at once. This technique is called *projection pursuit* in general, the observation of the whole animation process especially is called "grand tour" and may be rather time-consuming.

In every case data analysis helps to come to reliable data and to first ideas on its structure, especially on functional connections between the components of the data vector. The components are to be given as exact real numbers. Investigations of those connections considering also data uncertainty and allowing for stochastic or fuzzy connections can be found in Sects. 7.1 and 7.2.

2.2 Representation of Functional Relationships Basing on Data

As already announced, in this section data are considered as exact points, or, sometimes, as mathematical subjects (e.g. curves or surfaces) composed of them. Moreover, there will be no assumption made on how deviations may occur as will be done later on with respect to their assumed random character. The main task will be the mathematical representation of relations among the data by *functional relationships*, either only numerically or by formula expressions.

2.2.1 Relationships and Data

Statements about functional relationships are possible in different forms, as may be demonstrated by the following examples.

By *local interpolation* statements are made with respect to the numerical values of a functional relationship *within a neighbourhood* of a given observation point, *the datum.*

By a mostly only geometrical *global interpolation* a functional relationship will be given, which joins up the observed values, the *data*, over the whole interesting domain.

By *local approximation* an *approximate* representation of a functional relationship near a given datum is desired.

By *global approximation* an approximate representation of a functional relationship is to be computed, which is usable over the whole interesting domain.

For solving of all these tasks a rich variety of methods is provided in the usual software. The problem of the user, however, is again and again the question, *which* of the representation variants he should choose and *which* of the procedures offered he should use for the computation of the functional relationship in his given case.

The answers to these questions depend, as already discussed in the first chapter, essentially on the *information situation* given in the practical problem, and on the *aim* of the investigation intended.

The information situation is characterized by the following questions:

What is already known on the relationship, besides the given data? Is there any idea with respect to the possible *type* of the relationship, i.e. especially to the form of the curve (straight line, parabola, trigonometric function), exactly or approximately, from the theory of the branch of science or from experience? Or, if this is not the case, are at least some properties of the relationship known, e.g. with respect to its monotonicity or to its differentiability or to bounds, which cannot be crossed by it?

Moreover, the information situation includes the whole complex of *data quality*, which is excluded in this section by the assumption that all data are *reliable and exact*. Nevertheless sometimes this aspect will be considered, when the problems occur, whether deviations are neglectable, or random errors are possible or gross errors are to be feared.

Finally, an essential role for the information situation the fact will play, in which form the data are available. Obviously, there is a remarkable difference, whether the data are given by a *continuous* record of a function possibly superimposed by small errors or by observations at only few points. Cases, where the data are surrounded by an uncertainty part, randomly or fuzzily, will be considered later on in Chap. 7.

With respect to the aim of an investigation the main point of interest is *how the statement* about the relationship will be *used* in the practical problem.

The highest claim is meant, if the relationship obtained has to express the realization of a *technical regularity*. Here, absolutely, a *closed form expression* for the relationship will be necessary.

If a control process is to be started or supervised by the relationship, then a closed form is agreeable, but practically also a *pointwise representation* is sufficient for these purposes.

If only a prediction for a given part of the interesting domain is wanted, then an *approximate* representation of the relationship may satisfy.

Finally, if by investigation of the relationship only an *optimal value* of it should be found, then an approximate representation only within a neighbourhood of this optimum will be of interest.

Hence, information situation and aim control in common the choice of an appropriate *solving procedure*. In the given special case of determination of a functional relationship such a procedure uses an idea of a *type of dependence* and an *approximation principle*. Both should be chosen from practical considerations. Examples will be presented in the next subsections.

2.2.2 Interpolation

Now the simplest case is handled that the data are given only at *few* points and *without* any error. For the relationship wanted a *continuous* function is to be computed, which is, moreover, *locally monotonous*, i.e. between neighbouring data there are neither maxima nor minima.

As the aim of the task an approximate prediction of the functional values *between* neighbouring data is requested.

Moreover, in this subsection it is assumed that the distances between the data points are *small*, and, for the next consideration, that the function is *differentiable*.

First, the problem of *interpolation* is considered.

Because of the preceding assumptions for a function in question, which maps the (one-dimensional) x-axis into the (one-dimensional) y-axis, one can use the well-known TAYLOR-formula

$$f(x) = f(x_0) + (x - x_0)f'(\xi); \qquad \xi \in [x_0, x] . \tag{2.1}$$

Because of the assumed smallness of the distances between every single couple of neighbouring points, one can put within each of this single spaces approximately

$$f'(\xi) = \text{const.} \tag{2.2}$$

That means, in each single of these gaps $f(x)$ can be approximated by a straight line. Practically, one can join up each two neighbouring points by a line, an operation, which can be executed by any graphic software. The use of the functional values on the lines as approximating values of the functional relationship for the data is called *linear interpolation*.

The resulting broken line as an approximating representation of the function may be rather busy and can show *corners* at the data, which are really forbidden by the assumption of differentiability and being possibly senseless in the practical context. However, as already mentioned, the broken line should give an only local prediction of the functional values and, obviously, only *approximately*.

If this approximation does not suffice for the precision of the prediction, then one can go over to a TAYLOR-formula of higher order, e.g. to

$$f(x) = f(x_0) + (x - x_0)f'(x_0) + (x - x_0)^2 \frac{f''(\xi)}{2} . \tag{2.3}$$

In this case the interpolating values between the data lie on a *parabola*, in general on a graph of a polynomial of higher order. The coefficients of the parabola, or the polynomial of higher order, respectively, are *not uniquely* determined by the two neighbouring points, which they have to hit, as it was the case with the straight line.

Hence there are two different ways for an *interpolation of higher order*.

On the *first* way each time *several neighbouring points* are used for the interpolation.

For example the *cubic* interpolation is considered. In this case the desired curve should be interpolated between the data within the interval $[x_2, x_3]$ by a cubical parabola

$$a_0 + a_1 x + a_2 x^2 + a_3 x^3 = p_3(x) . \tag{2.4}$$

For a *unique* determination of $p_3(x)$ two further neighbouring points x_1, x_4 will be included and $p_3(x)$ has to hit all the *four* data points $(x_1, y_1); (x_2, y_2); (x_3, y_3); (x_4, y_4)$. This yields four equations with four unknowns, which are, in general, solvable uniquely:

$$\begin{aligned} y_1 &= a_0 + a_1 x_1 + a_2 x_1^2 + a_3 x_1^3 \\ y_2 &= a_0 + a_1 x_2 + a_2 x_2^2 + a_3 x_2^3 \\ y_3 &= a_0 + a_1 x_3 + a_2 x_3^2 + a_3 x_3^3 \\ y_4 &= a_0 + a_1 x_4 + a_2 x_4^2 + a_3 x_4^3 . \end{aligned} \tag{2.5}$$

The interpolation takes place only in the *middle* interval. By including of the additional neighbouring points the precision in the prediction interval will be essentially increased, in general. Numerical investigation on the real approximation precision within the prediction interval and in the given case needs reliable information on the real behaviour of the function to be interpolated and its derivatives and can be handled, e.g., by methods of interval mathematics (see Subsect. 3.1.2).

On the *second* way one will put a polynomial of higher order over each interval and will remove the ambiguity of the coefficients by demanding smoothness of the curve, when crossing from an interval to a neighbouring one.

Again the three intervals are considered: $\Delta_1 = [x_1, x_2]; \Delta_2 = [x_2, x_3]; \Delta_3 = [x_3, x_4]$. Over each interval a cubical parabola is defined:

$$\Delta_1 : y = a_{01} + a_{11}x + a_{21}x^2 + a_{31}x^3$$
$$\Delta_2 : y = a_{02} + a_{12}x + a_{22}x^2 + a_{32}x^3 \qquad (2.6)$$
$$\Delta_3 : y = a_{03} + a_{13}x + a_{23}x^2 + a_{33}x^3 .$$

This represents a system of three equations for the 12 coefficients. The demand that neighbouring curves should connect each other continuously at (x_2, y_2) and (x_3, y_3) does not yet fix the coefficients. Only the further demand that the crossing should be also smoothly, i.e. the corresponding values of the first and second derivatives of the neighbouring curves should be equal, yields the desired unique solution of the system.

This *spline* technique can be extended to arbitrary many intervals, for which procedure the continuity of the crossing at *all* the single data points is demanded (if using cubical parabolae up to the second derivative). This technique reminds of pliable curve templates, by which smooth curves can be drawn through a few single points. This task is now taken over by an appropriate software, which presents the result on the screen.

The interpolation problem is essentially more difficult, if the interpolation has to take place in a space of higher dimension. In the three-dimensional space the surface between three neighbouring points can be interpolated in a linear manner by a piece of a plane, and the different pieces will meet each other continuously. An *interpolation* with surfaces of higher order, however, will make sense only locally. So the non-linear interpolation can be modified by using *more* points from the neighbourhood of a "mesh", where the interpolation should be used, to determine the coefficients of the interpolating polynomials, though the expression obtained will be applied only *within* this mesh.

If the demands on splines for smooth crossing are transfered to a multidimensional space, then they must be valid for multidimensional subjects (curves, hyper-surfaces). Those demands can only be fulfilled for very special combinations of given mesh shapes and approximating functions in the single meshes. This problem will be of interest in connection with an approximation of functions of several variables, which have to satisfy a given mixed boundary problem, within the framework of the *method of finite elements* (FEM), as it is called (see some textbook on this topic, e.g. HUGHES (1987), and in Subsect. 2.4.2).

2.3 Local Approximation

If the formula expression for the desired approximate functional relationship should remain transparent and handy, then *the demand must be given up* that the approximating function has to *hit* all the data points *strictly*. This leads to a task, which is called *approximation* in a general sense.

2.3.1 Approximation

The *information situation* is now characterized by the allowance that the functional values y_i of the data may be superimposed by *small* not necessarily statistical *errors*, and by the assumption that *inaccuracy* with the corresponding argument values x_i may be *neglected*.

The *aim* is still the approximate representation of the unknown functional relationship, which the data fulfil, by a formula expression.

In contrast to interpolation, with which the local character is immanent, here *local* and *global* approximation are to be distinguished.

As with interpolation also with *local* approximation a contextual interpretation of the approximating function obtained is not intended. Only a representation of the graph of this function within a bounded domain is requested, from which approximate values can be read and predictions can be made.

As with interpolation the argumentation starts with the TAYLOR-formula and approximation is tried by *polynomials*.

In the following in this chapter only the simplest case is considered that the argument values x are taken from the one-dimensional space. The tranfer to the multidimensional space for the arguments does not offer any difficulty, in principle, but it is essentially more space-consuming in its presentation.

Hence now the data consist of the points (x_i, y_i), where $i = 1, \ldots, n$. For approximation a polynomial of order m

$$f_m(x) = \sum_{j=0}^{m} a_j x^j \tag{2.7}$$

will be used.

The central question of *every* local or global approximation is, how should the possible deviations of the approximating function $f_m(x_i)$ from the observed or measured data values y_i be *evaluated*.

According to a suggestion by GAUSS as a measure of deviation the square of the difference is chosen, i.e. $(y_i - f_m(x_i))^2$. This choice leads to simply manageable numerical tasks und will be used advantageously within mathematical statistics later on. Hence this measure can be never negative, the deviation of all data in common can be evaluated by the *sum*:

$$Q^*(y_1, \ldots, y_n; f_m(x_1), \ldots, f_m(x_n)) = \sum_{i=1}^{n} (y_i - f_m(x_i))^2 . \tag{2.8}$$

With all (x_i, y_i) given, Q^* is a function of the coefficients of $f_m(x)$:

$$Q(a_1, \ldots, a_m) = \sum_{i=1}^{n} (y_i - \sum_{j=0}^{m} a_j x^j)^2 . \tag{2.9}$$

The *best* approximation of the given data according to this principle is obtained by minimization of Q with respect to the coefficients $a_1, \ldots, a_m$.

This is a simple optimization problem, which can be reduced by equating to zero of the partial derivatives of Q into a linear equation system:

$$\sum_{i=1}^{n}\sum_{j=0}^{m} a_j x_i^j x_i^k = \sum_{i=1}^{n} y_i x_i^k \; ; \qquad k = 0,\ldots,m \; . \qquad (2.10)$$

Although the solution $(\hat{a}_0,\ldots,\hat{a}_m)$ can be given in a closed form (which will be presented and considered in Subsect. 7.1.2) at this point it is assumed for the moment that the solution is given by a computer output.

Hence the approximation of the unknown functional relationship $y = f(x)$ is presented as

$$\hat{y} = \hat{f}_m(x) = \sum_{j=0}^{m} \hat{a}_j x^j \qquad (2.11)$$

With that the considerations are mostly finished and the unknown function is *identified* with the obtained polynomial approximation. This *thoughtless* and *dangerous* inference, however, needs some *critical* assessment:

By the given procedure the deviations between the functional values and the approximating values become known *only* at the observation points, the data. In the spaces *between* these observation points there is *not any* information available on those deviations (but even here it will be of interest!). To have some confidence in these values in between it must be assumed that the functional values lie near each other and the unknown function is sufficiently smooth. The latter is the case, e.g., if it is known or may be assumed from the practical context that the unknown function has continuous derivatives of some higher order. The *ideal situation* is met, if it is known a-priori that the function itself is approximately a polynomial of some order m. Also here the argument with the TAYLOR-formula will help.

A further problem can be raised by the *numerical* solution of the system of (2.10), especially then, if many unknown coefficients a_j are to be determined simultaneously, that is the rule especially in the multidimensional case. In the following the numerical problem will be illustrated by a radically simplified demonstration example.

Let us consider the system of linear equations

$$1,00 a_0 + 1,00 a_1 = 1,00 \qquad (2.12)$$
$$1,01 a_0 + 1,02 a_1 = 1,03$$

with the obvious solution $(-1;2)$. An insignificant change of the coefficients, e.g. by an increase of the number of digits (notice that the preceding system arises from the following one by rounding of the coefficients), leads to the system

$$1,000 a_0 + 1,000 a_1 = 1,000 \qquad (2.13)$$
$$1,013 a_0 + 1,016 a_1 = 1,028$$

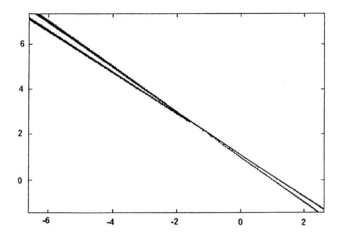

Fig. 2.2. A glancing intersection of two straight lines in the plane

with the obvious solution $(4; -5)$. Thus an insignificant changing of the co-efficients causes an enormous changing of the solution. This effect is called *numerical instability.* In the given case the reason for this phenomenon can be explained simply geometrically: The solution of a system of two linear equations with two unknowns is equivalent with the determination of the intersection point of two straight lines in the plane. If the two lines are *nearly parallel*, then they have a *glancing intersection.*

A small motion of the lines (i.e. a change of their coefficients) leads to an enormous moving of their intersection point.

This effect cannot be eliminated by choosing another solving procedure, even if this will seemingly yield a higher precision. The reason lies here in the choice of the *structure of the system of approximating polynomials.* To see this one should consider the behaviour of the basing functions

$$1, x, x^2, \ldots, x^m , \tag{2.14}$$

which will be linearly combined to a polynomial $f_m(x)$. Their graphs over the interval $[-1, 1]$ are shown in the following Fig. 2.3: As can be seen the curves are "nearly parallel" within $[0, 1]$. If the data lie within this part of the interval, then a numerical instability can almost be expected.

Hence sometimes it is suggested, especially if the argument values x_i will be needed as "standard argument values" for the observations, instead of the ordinary power functions, to use certain linear combinations $f_m^{(r)}(x); r = 0, \ldots, m$ of them, which are *orthogonal* over the set of arguments, i.e. having the property

$$\sum_{i=1}^{n} f_m^{(r)}(x_i) f_m^{(s)}(x_i) = 0 ; \qquad r \neq s \in \{0, \ldots, m\} . \tag{2.15}$$

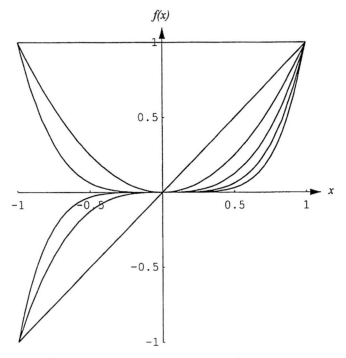

Fig. 2.3. Graphs of the functions x^j over $[-1,1]$

With these functions the solution of the system of linear (2.10) is trivial and the mentioned disagreeable effects cannot occur.

The same idea may be used, if the *argument values* x_i are to be determined beforehand, at which the function $y = f(x)$ is to be observed. According to a corresponding formula as (2.15) the ordinary polynomials $f_r(x) = x^r; r = 0, \ldots, m$ are *orthogonalized* with respect to the intended observation points x_i. The choice of those arguments is called orthogonal design of experiments and will be reconsidered in connection with regression theory in Subsect. 7.1.2.

Local approximations can be linked together as it was done by the spline technique with interpolations. In different fields various systems of approximating polynomials are used. In the one-dimensional case conditions for the smoothness of the crossings can be put, in the multidimensional case this leads to certain complications. Suggestions, how to overcome these problems, are treated in Subsect. 2.4.2.

Besides the ordinary polynomials and their multidimensional correspondences also other systems of functions are possible and useful as systems of approximating functions. Thus periodical functions are chosen, if it is known from the practical context that the unknown function will behave periodically. If, e.g., it is known that the original period of this function is q , i.e. $f(t + q) = f(t)$, then by the scale transformation $x = (2\pi/q)t$ and with the system of functions

$$1, \cos x, \sin x, \cos(2x), \sin(2x), \dots \qquad (2.16)$$

one arrives at the FOURIER series setup

$$f_{Fm}(x) = a_0/2 + \sum_{j=1}^{m/2} a_j cos(jx) + \sum_{j=1}^{m/2} b_j sin(jx) , \qquad (2.17)$$

as it is called. Other representations are possible and usual. Likewise there are corresponding setups for the multidimensional case. For the numerical determination of the coefficients $a_0, a_j, b_j; j = 1, \dots, m/2$ appropriate software exists, at least for the usually equidistant observation points x_i. With respect to a closed form solution of the resulting systems of equations see some textbook of this topic.

A modification of the approach is necessary for the case that *the datum itself is a function* or some hypersurface. This occurs, if the data values are obtained continuously, e.g. measured by electronic means. Again only the one-dimensional case is considered that the datum is given *geometrically* or pointwise with neglegible spaces between, as a curve $y = g(x)$. The inaccuracy of the measurements of x and y may be neglegible. This characterizes the given *information situation.*

The *aim* is a *local* approximation by a *closed form* function, an *analytical expression* as it is called. The motives for this aim may be diverse. It may simply be the desire for a data reduction, since the pointwise storing is too space-consuming. A *comparison with other functions* is most comfortable, if these are given in the same form with corresponding coefficients being synonymous with each other. But also it may be the request of only a reconstruction of the function when required.

As with approximation on the basis of a few discrete points an appropriate system of functions is to be chosen, polynomials, trigonometrical or other functions.

The *approximation in the quadratic mean* over the interesting *local* domain B is effected by minimizing the integral

$$Q(a_1, \dots, a_m) = \int_B (g(x) - \sum_{j=1}^{m} a_j f^{(j)}(x))^2 dx , \qquad (2.18)$$

with respect to the arguments, at it was done in the case with discrete points. By equating to zero of the partial derivatives of Q with respect to the a_j again a system of linear equations, now

$$\sum_{j=1}^{m} a_j \int_B f^{(j)}(x) f^{(l)}(x) dx = \int_B g(x) f^{(l)}(x) dx , \qquad (2.19)$$

with $l = 1, \ldots, m$ is obtained, the solution $\hat{a}_1, \ldots, \hat{a}_m$ of which can be given numerically. For a unique solution of (2.19) it is necessary that the functions $f^{(l)}(x)$ are linearly independent, i.e. the following equation

$$\sum_{j=1}^{m} a_j f^{(j)}(x) = 0 \qquad (2.20)$$

has, demanded for all $x \in B$, only the trivial solution $a_1 = \cdots = a_m = 0$. If there would exist a *non-trivial* solution, then at least one function could be expressed by a linear combination of the others and hence would be superfluous.

Also in the case of continuous data the problem of numerical instability can occur. Hence also here it is suggested to use systems of functions, which are *orthogonal* over the domain intended for approximation, i.e. systems, which fulfil the condition

$$\int_B f^{(j)}(x) f^{(l)}(x) \mathrm{d}x = 0; \qquad j \neq l \in \{1, \ldots, m\} \ . \qquad (2.21)$$

By successive application of these conditions a given linearly independent system of functions can be orthogonalized. As an example the system of polynomials (2.14) with $m = 5 : 1, x, x^2, \ldots, x^5$ over the domain $[-1, 1]$ is considered, which is shown in Fig. 2.3. Every finite interval can be normalized to this interval. The result of that orthogonalization are the LEGENDRE polynomials:

$$
\begin{aligned}
f^{(1)}(x) &= 1 \\
f^{(2)}(x) &= x \\
f^{(3)}(x) &= \frac{1}{2}(3x^2 - 1) \\
f^{(4)}(x) &= \frac{1}{2}(5x^3 - 3x) \\
f^{(5)}(x) &= \frac{1}{8}(35x^4 - 30x^2 + 3) \\
f^{(6)}(x) &= \frac{1}{8}(63x^5 - 70x^3 + 15x) \ ,
\end{aligned}
\qquad (2.22)
$$

which are presented in Fig. 2.4, which may be compared with Fig. 2.3. By using of orthogonal functions the solution of the corresponding systems of equations is trivial, because only the diagonal elements of the system matrix are different from zero. Together with the constants

$$\int_B f_j^2(x) \mathrm{d}x = \gamma_j \ , \qquad (2.23)$$

to be determined beforehand, the minimal quadratic deviation Q_{min} is obtained by

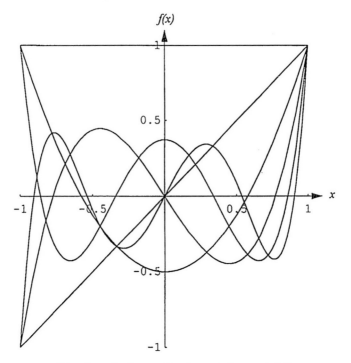

Fig. 2.4. Orthogonal polynomials over [−1,1]

$$Q_{min} = \int_B g^2(x)\mathrm{d}x - \sum_{j=1}^m \hat{a}_j^2 \gamma_j \ . \tag{2.24}$$

From this result some further *advantages* by the application of orthogonal setups can be seen. Every inclusion of a further orthogonal function improves the approximation. This is valid naturally also for non-orthogonal systems, but it is recognized not so clearly there. For orthogonal systems, however, the coefficients $\hat{a}_j$ need not be computed anew for the previous functions, when a next function is included, because they remain the same ones.

A possible *disadvantage* of orthogonal systems is, in general, that the single orthogonal functions cannot be *interpreted* so simply in the practical context.

The transfer to *multidimensional* orthogonal systems is without any problem, in principle, but it is essentially space-consuming and sometimes confusing. The construction of those systems may be effected by tools for formula manipulation (for example: Mathematica).

2.3.2 Other Approximation Principles

The approximation principle "in the quadratic mean" (see (2.18)), as presented up to now, *tolerates very large deviations* (peaks), if only the set, on which they occur, is sufficiently "small". That may be undesirable, perhaps even dangerous, because possibly already a short-term leaving of the "security belt" may cause a damage.

Hence also other approximation principles as that suggested by GAUSS are of practical interest.

Such a principle, named after TSCHEBYSCHEFF, demands to minimize the *largest absolute* deviation by the approximating function. In the continuous case this means for the expression

$$Q_T(a_1, \ldots, a_m) = \max_{x \in B} \left| g(x) - \sum_{j=1}^{m} a_j f_j(x) \right| \tag{2.25}$$

that it is to be minimized by determining suitable a_j. For this *uniform approximation* a far reaching theorem is valid that every continuous function can be approximated by polynomials arbitrarily precisely. This statement makes the use of polynomials attractive for a uniform approximation.

With the *pleasant* property that the tolerance against short-time peaks is prevented by the uniform approximation some *unpleasant* properties are connected, which one has to weigh up the one and the others in a concrete case. The approximation is assessed according to the *most disadvantageous* point, which may, however, be practically without any importance. This fact may be considered, e.g., if there is a chance for the occurence of undetected outliers. Moreover, for this type of approximation there is not so a nice simple mathematics, which is transparent also for laymen. But this will be of decreasing importance by the increasing efficiency of computers and numerical approximating procedures. Especially in the case of multidimensional domains B the computation of the parameters $\hat{a}_j$ can be rather time-consuming.

In the case of a few *discrete* data (x_i, y_i) the Q_T has the analogous form

$$Q_{Td}(a_1, \ldots, a_m) = \max_i \left| y_i - \sum_{j=1}^{m} a_j f_j(x_i) \right|. \tag{2.26}$$

The solution of the corresponding minimizing problem with respect to the a_j will be handled by routine procedures of linear programming.

Also here again the comment is useful that about the approximation *between* the observation points, in general, *nothing* can be stated, except that information are given about the behaviour of the unknown function in the whole domain of approximation.

If only the *quadratic* weighting of the deviations is felt unreasonable (e.g. as too strong), then for the approximation the principle of the *smallest sum of the absolute values* may be chosen. Then the functional

$$Q_{MKA}(a_1, \ldots, a_m) = \int_B \left| g(x) - \sum_{j=1}^{m} a_j f_j(x) \right| \mathrm{d}x \tag{2.27}$$

is to be minimized by a suitable determination of the a_j. Also here a simple and transparent procedure for a closed form solution is lacking. Again approximating optimizing procedures are to be applied. The analogon for the case that only a few discrete data are given has the form

$$Q_{MKAd}(a_1,\dots,a_m) = \sum_{i=1}^{n} \left| y_i - \sum_{j=1}^{m} a_j f_j(x_i) \right|. \qquad (2.28)$$

Also in this case for the determination of the optimal coefficients $\hat{a}_j$ methods from linear programming are used. For an assessment of the approximation *between* the points the same comment is given as for the principles already mentioned for this case.

Theoretically, the power exponent p, by which the deviations are weighted ($p = 2$: quadratic mean; $p = 1$: absolute value) can be put to any positive value; even the TSCHEBYSCHEFF-approximation can be included in this scale by taking p "infinitely large".

The principles show a different sensitivity against untypical data, e.g. against outliers. This may be important in certain cases. Thus this sensitivity *increases* from the principle of the least absolute values over the GAUSS-principle to the TSCHEBYSCHEFF-principle.

2.3.3 Empirical Smoothing

The presented approximation principles can also be used for a *local smoothing* of a given data material, given as finite sets of data points or as curves or surfaces.

In contrast with the preceding subsections *no* representation by mathematical formulae is demanded; only a smoothed geometrical representation is requested, which the relationship may change only insignificantly within a small local domain. Hence the representation should remain *data oriented*. Assumptions on the form of the relationship are not (yet) available, one hopes to obtain a hint to a possibly suitable function by this smoothed representation.

With respect to the *information situation* nothing is known besides the data, which are assumed to be obtained *exactly*. It is only presumed that the data could be superimposed by small deviations or errors, on the character and extent of which no assumption can be formulated though. The *aim* is only the construction of a "smoothed picture" and the hope for an idea to grasp it by a function. This aim is typical for an early stage of the investigation. Since there is no information besides the data, a larger size of them is necessary (Golden rule).

The procedure, now to be presented, is sometimes called *empirical regression*, because it is frequently used in preparation of a statistical treatment of the data.

First a *window* is to be chosen, as it is called, i.e. a small domain, from which data will be used for the smoothing. Two types are distinguished: a "hard" window, a domain of simple shape, an interval in the one-dimensional case, and a "soft" window, such a domain endowed with a weighting function $h(x)$ defined over it, which assumes its maximum value 1 somewhere in the "centre" of the domain and with non-negative values monotonously decreasing towards the borders.

Then the type of function is to be fixed, with which the local smoothing is to be effected. Usually these are a constant $f(x) = a_0$; a linear function, in the one-dimensional space $f(x) = a_0 + a_1 x$; and at most a quadratic function, correspondingly $f(x) = a_0 + a_1 x + a_2 x^2$.

Finally, an approximation principle is to be chosen, e.g. the sum of the squared deviations as the criterion.

After these choices of the three components the procedure begins in a certain starting window F_1. As an example the following one is chosen: A soft window with the weighting function $h(x)$, a linear approximating function $f(x) = a_0 + a_1 x$, and the method of least sqares.

In F_1 the coefficients a_0 and a_1 are computed using the weight $h(x_i)$ for each datum (x_i, y_i) according to

$$\sum_{i=1}^{N} h(x_i)(y_i - a_0 - a_1 x_i)^2 = \min . \tag{2.29}$$

Note that the sum in fact runs only over such data, which are in the window F_1, because the function $h(x)$ is equal to zero outside. The function $\hat{f}_1(x) = \hat{a}_0 + \hat{a}_1 x$ formed with the solution $\hat{a}_0$ and $\hat{a}_1$ of (2.29) is used for approximation only in the centre (and a small neighbourhood) of the window. Then the window is shifted by one half of this small neighbourhood and denoted by F_2. Then the procedure is repeated until the whole interesting domain is succesively covered over. Because this procedure makes sense, practically, only at the screen, where the result is presented, the "granularity" of the screen effects that the single small shiftings are no longer perceived and the impression of a continuous, even smooth approximation function arises.

2.4 Global Approximation

Within *global approximation* two directions can be distinguished.

In the first case from the given data an expression in a closed form is to be derived being valid for the whole interesting domain and containing parameters, the number of which being as small as possible, but nevertheless representing the unknown function sufficiently precisely. Here the practical main problem consists in the choice or the bringing out of a *setup* , i.e. of a mathematical structure for the approximating function.

In the second case the unknown function is "known" *implicitly*, e.g. as the solution of a given mixed boundary value problem, or even *explicitly*, e.g. as an empirically given curve or surface, e.g. recorded by an electronic medium.

This function is to be represented by a series of local standard functions (i.e. functions, which are different from zero only over a small subdomain), in order that the specialist will be able to consider or investigate the local behaviour of the function at every place of the interesting domain. Examples of this case are supplied by *the method of finite elements* and by the conception of the *wavelets*.

2.4.1 Approximating Functional Relationships

In the present context a *functional relationship* is understood as a relation between one (or several) dependent variable and one (or several) explanatory variable, which should be represented by an expression in a closed form containing yet some unknown parameters, which may take values from a given index set. These parameter values are to be determined or computed e.g. from given data. For the sake of simplicity of presentation only the simplest case of *one* dependent variable is considered, thus the functional relation

$$y = \eta(x; a_1, a_2, \ldots, a_m); \qquad (a_1, a_2, \ldots, a_m) \in A . \qquad (2.30)$$

The function η, giving the structure of the mathematical connection up to parameters, is usually called *setup*. Its expertly supported choice is the *main problem* with global approximation.

Global approximation differs from local approximation, as is already mentioned in Sect. 2.3, only in the *aim*, the information situation is the same.

A scientific, technical or economic regularity is to be approximately put into a mathematical form, with which theoretical considerations as well as concrete predictions for yet unknown situations can be effected. The number of parameters should be as small as possible, to guarantee tranparency and easy handling.

Starting point for the *setup choice*, as it is called, should always be theoretical considerations on the situation, which is to be described, and examination and reflection of already available numerical results, in doing so a sound scepticism is advised. All which seems interesting and important for the case given should be kept in mind, let it be only for later assessments or improvements of the results obtained. These preliminary considerations should end with a supposition of the form of the relationship, i.e. with a setup.

It is advantageous in every case to have a setup supported by the field of application, even if therein some unknown parameters occur in a *non-linear* manner, as e.g. in

$$y = a_1 + a_2 \exp(a_3 x^{a_4}) , \qquad (2.31)$$

which will be used in an example in Subsect. 7.2.2. Such setups have the great advantage that they can be interpreted in the practical context and hence are comprehensible in their meaningfulness.

In case that such an expertly supported setup is *not* (yet) available or that the unknown function can only be given as a numerical solution of some mathematically complicated problem, e.g. a differential equation, then one should be content with an *approximating setup* using a system of functions (polynomials, trigonometrical or exponential functions), in the choice of such a system taking into consideration known properties of the unknown function.

In both cases it can occur that several setups seeming to be suitable and, so to speak, compete for the title "best setup". Here a criterion is to be chosen to decide on the quality of the setup, e.g. the value of the criterion for approximating the given data. This problem is considered in Subsect. 7.1.3 in connection with methods of mathematical statistics.

On no account the setup choice should be left to a software tool offering a catalogue of function types, which may be combined arbitrarily to a "setup". For this "setup" the computer supplies an approximating function, which approximates the given data "in the best way" according to the given criterion. Such an approach leads to a result, which needs neither be interpretable in the given context nor be *stable* with respect to the data, i.e. slightly changed or further data can result in a possibly completely other selection from the catalogue.

The method for determination of a global approximating function numerically is totally analogous to that used for local approximation. After the choice of an approximation criterion such parameter values $\hat{a}_1, \hat{a}_2, \ldots, \hat{a}_m$ are computed, which minimize it. The only difference consists in the fact that now the whole interesting domain G and *all* the given data are included simultaneously.

Linking together of local approximating functions over the whole domain constitutes *no* global approximation according to the explanation given above, since it does not justice to the chosen aim for that.

Though it can occur that the problem itself suggests the use of *different* setups in *different* subdomains of G. It can happen, however, that e.g. turn-over points, boundary surfaces of layers, take-over times, or similar factual changings divide the domain G into *phases*, as they are called. Taking into account only this partition by different setups in different subdomains does not lead away from global approximation. An interesting mathematical problem arises if the turn-over points or other phase borders are *unknown* and should be determined from the data at the same time. This problem is subject of *multi-phase regression* (see some textbook on this topic and for an example BANDEMER/SCHULZE (1985)).

2.4.2 Approximation with Locally Variable Setups

As already mentioned in the preceding subsection the concept of global approximation can also be used sensibly, if a mathematical model of the situation is given, e.g. by expertly supported considerations, but its adaption to the given data consists in a complicated numerical problem, for which a solution in a closed form does not exist.

Typical cases for this problem area are mathematical models in form of *differential equations* (mixed boundary value problems) and *integral equations*. Since the specification of such equations is always effected by a process of abstraction and idealization, a *highly precise* solution of the arising mathematical tasks would be senseless for practical requirements, even if the data used are assumed to be given exactly, which is, as is well known, also an idealization.

Since the field of numerical handling of such equations is very wide (see some textbook on this topic) the treatment of this field in the present context will be restricted to a short consideration of simple mixed boundary value problems, in order to fit a modern concept into the framework of this section.

The approximately solving of such problems usually starts with a setup of the type

$$u(x,y) = \varphi_0(x,y) + \sum_{k=1}^{m} c_k \varphi_k(x,y) \,, \tag{2.32}$$

where the function φ_0 has to satisfy the non-homogeneous boundary conditions, whereas the φ_k fullfil the homogeneous ones. (RITZ-setup). This setup is then introduced into a certain functional, for which the stationary points with respect to the coefficients c_k are computed. But this well-known procedure is practicable only for "nicely regular" domains G. Hence a method due to GALERKIN is preferred, in which the setup mentioned above is introduced into the differential equation and the deviations obtained, remainders $R(x,y)$, as they are called, should be minimized by the choice of the c_k. Then an approximation functional is chosen, being quadratic in the coefficients. This leads, in analogy to the method of least squares to the system of equations

$$\int_G R(x,y)\varphi_j(x,y)\mathrm{d}x\,\mathrm{d}y = 0; \quad j = 1,\ldots,m \,. \tag{2.33}$$

With very complicated regions G, however, it is *hopeless* to find functions φ_k, which satify the demanded conditions in the whole domain. Therefore the *method of finite elements* (for details see some textbook on this topic, e.g. HUGHES (1987)), was developed, with which approximating solutions can be obtained for practically interesting but rather irregularly shaped domains.

In this method first the domain G is split up into *elements*, as they are called. These elements are relatively small, up to their boundaries disjoint subdomains G_j, by which the whole domain is exhausted. In each single of the G_j the unknown function $u(x,y)$ is approximated by a problem-appropriate but simple setup. Usually simple polynomials at most of third order in each variable are used. To guarantee continuous crossings between the elements, the form of the polynoms in neighbouring elements must be suited to each other. If this proves impossible, sometimes continuity of the crossing is renounced.

To fulfil the conditions of continuity in a practically easily comprehensible manner setups of the form (2.32) with their coefficients c_k are *unsuitable*. Rather a *representation of such functions* in the single elements should be looked for, in which special values of the unknown function and possibly of its (partial) derivatives in certain points in the elements occur. These points are called *nodal points* and their values are considered as "free parameters", *nodal variables* as they are called. These coefficients, denoted by $u_i^{(l)}$ and later also by u_k may also be *sometimes values of derivatives*. Instead of the simple powers in setup (2.32) special polynomials are introduced, now called *shape functions*.

As an example let be considered the p nodal variables $u_i^{(l)}$ at the nodal points $(x_i^{(l)}, y_i^{(l)})$ in the l-th element G_l, then one gets the representation of the unknown function $u^{(l)}$ in this element

$$u^{(l)}(x, y) = \sum_{i=1}^{p} u_i^{(l)} N_i^{(l)}(x, y) \tag{2.34}$$

with shape functions $N_i^{(l)}$, each one only defined over G_l, for which functions a sort of orthogonality should be valid:

$$N_i^{(l)}(x_j^{(l)}, y_j^{(l)}) = \begin{cases} 1 & \text{for} \quad j = i , \\ 0 & \text{for} \quad j \neq i . \end{cases} \tag{2.35}$$

Since the nodal points very frequently are put on the boundaries of the elements, in neighbouring elements also nodal variables occur, which are the same. Taking this into account with (2.34) a setup can be generated over the whole domain G by combining and renumbering:

$$u(x, y) = \sum_{k=1}^{n} u_k N_k(x, y) . \tag{2.36}$$

This setup corresponds, with respect to its intention, to the approach with the RITZ-setup; but now the coefficients u_k are, at the same time, the function values of the requested function at the nodal points, and the setup function $N_k(x, y)$ have each only a local support. In the setup (2.36), which represents the *main formula* of the method of finite elements, the geometrical boundary conditions can be taken into account in a simple manner, by putting the corresponding nodal variables to the given values.

By inserting the whole setup into a corresponding (quadratic in u_k) functional and demanding its stationarity a system of linear equations for these values of the function arises. In this way the mixed boundary value problem is (approximately) reduced to a system of linear equations with very many unknowns. Since always only values of the function from neighbouring elements will occur at the same time, the system matrix $\mathbf{S}$ of the system of equations has a rather special structure with many vanishing elements (sparse matrix), which makes the solving of the system possible in spite of the high number of unknowns.

Naturally, also the GALERKIN-approach can be used as an optimizing functional for a minimizing of the residua.

A presentation of the internal solving procedure can be omitted here, since a treatment of the task is practicable only by means of suitable hard- and software. With respect to further details see a corresponding textbook, e.g. HUGHES (1987).

The problem is interesting for data analysis by the influence of *model impreciseness*. The mixed boundary value problem is an idealization of the practical problem; hence the coefficients of the differential equation may be characterizing values of the material used, which are known, in the case given, only *imprecisely*. Moreover, even with the initial and boundary values impreciseness can occur, e.g. when they are obtained by measurement or observation. Both influences are investigated today by sensitivity analysis, by which the effect of small changes is considered theoretically. An assessment of the magnitude in the concrete situation is seldom effected. The problem of impreciseness, however, is of great importance for the choice of a meaningful numerical preciseness for the procedure of the method of finite elements. Mostly things are overdone here and a precision is demanded, which suggests a real precision of the solution for the problem at hand proving *totally virtual*, if the impreciseness of the model and of the initial and boundary values is taken into account.

A well known method for an approximate representation of a graphically given periodical function is its expansion in a FOURIER series . The *approximation* is effected by using only some terms from the beginning of the series, such that the graph of the function and the graph of its reconstruction by this partial series can not be distinguished at first glance. This approach was improved extensively and in different directions. The function need no longer be *periodical* (the period is "infinitely long"). For this case (orthogonal) systems of functions are developed, in which each single function is different from zero only on a finite interval, which decreases in length with increasing index of the function. This approach was generalized in different respects. For a comprehension of this direction of development the consideration of a suggestion by HAAS (1910) will suffice, called already classical today. He considered the function

$$\psi(x) = \begin{cases} 1 & \text{for} \quad x \in [0, 1/2] , \\ -1 & \text{for} \quad x \in (1/2, 1] , \end{cases} \tag{2.37}$$

and formed with it

$$\psi_k^j(x) = 2^{j/2} \psi(2^j x - k) ; \quad j, k = 0, \pm 1, \pm 2, \dots . \tag{2.38}$$

The ψ_k^j are different from zero exactly over the interval $[2^{-j}k, 2^{-j}(k+1)]$. This system of functions forms a basis for the set of all twice continuously differentiable functions over the x-axis. With functions of this kind, but offering a richer set of functional values, predominate frequencies *and* abnormal amplitudes can be recognized rather well *at the same time*. Such systems of functions are called *wavelets* and used successfully for different practical purposes, e.g. for data compression, where only a few coefficients of the expansion are stored, from which the original function can be reconstructed with sufficient precision (see some textbook on this topic, e.g. WALNUT (2002)). Also here the consideration starts with the assumption that all original data are exact and all pecularities found are of equal interest. To this day the real assessment of the necessary precision for a given application purpose is an open problem. For investgation of the properties of a wavelet usually the starting point are *test functions*, which show certain phenomena, as they may also occur in the planned field of application, and from the results for these test functions the properties of the wavelet are concluded. The use of wavelets makes sense only with an *extensive* data material, as it is given e.g. with grey tone pictures on pixel level.

2.4.3 Approximation in Differential and Integral Equations

A well-known field of application of global approximation is the determination of interesting constants in differential or integral equations. The information situation consists in the assumptions that, on the one hand, a *practical* solution of the equation is given, graphically or only in some discrete points, and, on the other hand, a solution in a closed form or an approximating solution is known represented by a functional relationship, which contains unknown parameters yet.

After the choice of an approximation principle, meaningfully the same as for the approximating solution, the unknown parameters are determined by minimization according to this principle. Also here the *real* approximation precision between the practical problem, the differential or integral equation, and the given practical solution, and hence the *real* precision of the determined parameter values, which is important for their further use, remain in darkness. In practical cases, however, this precision will be rather small, hence in many cases assumptions suggest theirselves that the parameter values may be realizations of random variables (see Subsect. 4.1.3).

2.5 Approximate Optimization of Empirical Functions

If, in the *information situation*, the values of the unknown function are available only pointwise (and perhaps only approximately) by measurement or observation, which causes expenditure of time and money, but if the *aim* is only the determination of the (rough) position of its optimum, then an *explicit representation* of the function, e.g. by a suitable setup, is *not necessary*.

For this problem different procedures are offered, the effectivity of which depends on the dimension of the space of arguments and on the assumption on the precision of the observation method to determine the "true" values of the functions.

If the dimension of the argument space is high, but at the same time the precision of the observations is high, then *search procedures* are appropriate.

In the *blind* search the choice of the argument points, where the observations should take place, follows the uniform distribution over the whole interesting domain G. The points are fixed by a random number generator according to the distribution: Every point has the same probability to be chosen.

In the *controlled* search the probability distribution over G is modified after each single observation. The search starts with a certain a-priori distribution, as it is called, around the presumed position of the optimum. This approach was developed further by the creation of *genetic algorithms*, as they are called, into a subtle heuristics, which claims to model evolutionary strategies of Nature. The procedures are rather effective for the case considered (see for details e.g. GOLDBERG (1989)).

If the dimension of the argument space is *moderate*, but remarkable observational errors are to be expected, then another sequential approach is recommended. In a (small) starting subdomain the function to be optimized is approximated by a hyper-plane and the coefficients of this hyper-plane are determined by approximation from observations in the subdomain. The position of the hyperplane supplies a rough direction of the tangential plane and hence the direction, in which the function changes its values most strongly. In this direction in increasing distances from the starting subdomain observations take place as long as remarkable changings (according to the interest – towards a minimum or a maximum) can be recorded. If the changings are too small or in the "wrong" direction, then a new "basing subdomain" is specified in the neighbourhood of the last observed points and the proceeding is repeated. In this way a subdomain is reached, where remarkable changings can no longer be observed in any direction. Then it is assumed that a neighbourhood of a stationary point is reached and the unknown function is approximated in that subdomain with a quadratic setup in order to locate the optimum further. This procedure was suggested by BOX/WILSON (1951) and uses a statistical background and methods of statistical experimental design for the choice of the observation points aiming at minimizing the observation costs in a wide sense.

3

Specification and Use of Observation Fuzziness

In this chapter impreciseness and vagueness of data (sometimes also of components of the model) are treated purely phenomenologically. Imprecision is considered given and tried to be specified mathematically. The data background is not analysed and not any assumption is put with respect to models for interpretation and for specification of the process of data genesis. Possibilities for such data models are introduced and treated in the following chapters, so, e.g., stochastical assumptions in Chap. 4.

3.1 Specification by Intervals

The mathematical form of specification and treatment of data imprecision depends essentially on how large imprecision of the values is when compared with the scale of the values and their variation. Hence at first the simplest case is considered, when that imprecision may be assumed to be *small* in comparison with the values.

3.1.1 Simple Error Propagation

An approach, which can be already called classical, for inclusion of impreciseness caused by measurement or observation is the consideration of *error propagation*.

Here the starting point is the information situation that the imprecision of certain argument variables x_j can be specified by *intervals*, in which the "true" values of the corresponding variables are situated with certainty. The generalization to multidimensional closed (convex) domains for several argument variables simultaneously makes sense theoretically, but will not be treated here.

Moreover, it is assumed that the *functional relationship*

$$y = f(x_1, \ldots, x_m) \tag{3.1}$$

is known accurately and surely, i.e. a possible model incertainty is neglected in the face of argument impreciseness.

The *aim* of investigation is an assessment of the imprecision of the value y of the function.

For that two further assumptions are introduced. On the one hand it is assumed that the function f is differentiable with respect to its arguments, and on the other hand that the imprecision of the measurements is *small* when compared with the variations of these derivatives.

On these conditions the TAYLOR-formula can be approximately used, written with the argument value intervals Δx_j in the form

$$\Delta y = \sum_{j=1}^{m} \frac{\partial f}{\partial x_j} \Delta x_j \qquad (3.2)$$

In this representation the partial derivatives $\frac{\partial f}{\partial x_j}$ are to be interpreted each as the mean value within the interval Δx_j. By this and by further neglections, inter alia with respect to the approximation of f the interval Δy is also only an approximation, of which one can no longer say that it contains the "true" value with certainty. Hence this approach is taken as a starting point for a *stochastic* idea of the observation imprecision, which leads then to the *error propagation law* due to GAUSS (see Subsect. 4.1.3). Because of the assumed relative smallness of the intervals for the variables the interval formula as above gives a sound impression of the scale of the imprecision of the function value y.

If the functional relationship has the special form of an exponential product

$$y = k x_1^{a_1} x_2^{a_2} \cdots x_m^{a_m} , \qquad (3.3)$$

which occurs frequently in application, by logarithmic differentiation, as it is called, an assessment of the *relative error* is obtained

$$\frac{\Delta y}{y} = \sum_{j=1}^{m} a_j \frac{\Delta x_j}{x_j} . \qquad (3.4)$$

Also starting from this expression a special form of the (stochastic) error propagation law is supplied.

3.1.2 Basic Ideas of Interval Mathematics

The formulae of error propagation give only an impression on the possible scale of deviations of the function values, when the scales of imprecision of the argument values are known. This problem can be sharpened to a mathematical task:

For the argument values x_j *exact* intervals $x_{ju} \leq x_j \leq x_{jo}, j = 1, \ldots, n$, shortly $[x_{ju}, x_{jo}]$, are given, which interval (as small as possible) includes then all possible values of the function $f : I\!R^n \to I\!R^1$ for argument values from those intervals? Obviously instead or besides these argument values also parameter values may be given as intervals, the influence of which on the function values is to be specified.

This task demands so the development and investigation of computing methods for intervals, an *interval mathematics*. But already an *interval arithmetics*, i.e. the treatment of basic arithmetical operations for intervals, shows some drastic differences to methods used for pure numbers. It seems reasonable to introduce the *resulting intervals* as sets of all possible results for single values:

$$A * B := \{a * b \mid a \in A, b \in B\} \quad \text{with} \quad * \in \{+, -, \cdot, :\} \ . \tag{3.5}$$

When applied to concrete intervals $A = [a_u, a_o], B = [b_u, b_o]$

$$A + B = [a_u + b_u, a_o + b_o] \ , \tag{3.6}$$

$$A - B = [a_u - b_o, a_o - b_u] \ , \tag{3.7}$$

$$A \cdot B = \Big[\min\{a_u b_u, a_u b_o, a_o b_u, a_o b_o\} \ , \tag{3.8}$$

$$\max\{a_u b_u, a_u b_o, a_o b_u, a_o b_o\} \Big] \ ,$$

$$A : B = [a_u, a_o] \cdot [1/b_o, 1/b_u]; \quad 0 \notin B \ , \tag{3.9}$$

already strange effects occur.

For $A = [-3, 1]$ and $B = [2, 4]$, e.g., this results in:

$$A - A = [-4, 4]; B - B = [-2, 2]; B : B = \left[\frac{1}{2}, 2\right] \ .$$

Addition and multiplication are still commutative and associative, i.e. one can change the sequence of summands or factors and put in brackets arbitrarily. But for a simultaneous use of addition and multiplication there are no longer a distributive law, by which sums could be multiplied out. It holds only

$$A(B + C) \subseteq A \cdot B + A \cdot C \ , \tag{3.10}$$

which leads, e.g., for $A = [-2, 1]$ to the astonishing result $A^2 + A = [-4, 5]$ and $A(A + 1) = [-4, 2]$. Hence it is reasonable to put in brackets as many as possible in arithmetical expressions to keep the result as narrow as possible. Since the difference of equal interval numbers is *no longer* the number 0 and the quotient of equal interval numbers is *no longer* the number 1, equations for intervals can *no longer* be transformed by subtraction or division applied to both sides.

When using interval matrices and interval vectors, i.e. matrices and vectors the elements of which are intervals each, one can form the product as usual, e.g. to represent a system of linear equations: $\mathbf{Ax} = \mathbf{b}$, where $\mathbf{A}$ is an interval matrix and $\mathbf{x}$ and $\mathbf{b}$ are interval vectors of appropriate orders. This form is then the starting point for the determination of an interval evaluation for the unknown interval vector $\mathbf{x}$, if $\mathbf{A}$ and $\mathbf{b}$ are given. The solving methods for that task are purely numerical procedures, by which always including sets for $\mathbf{x}$ are determined; *closed form solutions* can *not* be obtained by means of interval arithmetics. In the included single subprocedures, in order to secure an always sure inclusion of values, "rounding" is always effected outwardly, this must be guaranteed by a suitable software.

The demand that the including set for all possible solutions should always have the form of intervals leads in many cases to a *blowing out* of the requested supersets, which can make the final statement of the interval arithmetical treatment practically useless. This would be obvious especially if for the inverse $\mathbf{A}^{-1}$ of a matrix $\mathbf{A}$ with intervals as elements an inclusion by a matrix *of even this type* should be specified.

Trying to escape from this *overestimation effect* also *other types* of including sets (e.g. parallelepipeds, balls) are considered, or a suitable partition of the interesting domain in subdomains is introduced, in each single of which inclusions are given by intervals or vectors of intervals, respectively.

The search for methods to find including sets *as small as possible*, i.e. sets including really possible solving values and possibly only those, is an always topical field of research in interval mathematics.

The field of application of interval mathematics comprehends the whole numerical mathematics; determination of the zeros of a function, solution of systems of linear or nonlinear equations, computation for interpolation and approximation, linear and nonlinear programming, numerical treatment of mixed initial value problems for ordinary and partial differential equations and integral equations.

For details see e.g. MOORE (1979) and ALEFELD/HERZBERGER (1983).

In real situations in application some imbalances may occur in several respects. On the one hand the intervals for the input values should be specified as *precisely* as possible, which can be effected in practical cases only with some arbitrariness. Hence application of interval mathematics is recommended, as a rule, if the specified intervals are *small* when compared with the variability of the quantities considered, hoping that in such cases minor misspecifications remain without perceptible consequences. On the other hand interval mathematics demands for a *practically infinitely precise* performance of the chosen procedure, in order to include the changings of the intervals always towards the "safe" side. This requires a lot in the performance of the computations necessary. But this difficulty is defused in an increasing extent by the development of computer hardware. Finally by an *overemphasising* of specification and acknowledgement of imprecision in the arguments and parameters a *feeling of safety* is generated for the user, which cannot be supported by the knowledge on the determining role of the *weakest link* in the chaine: model – procedure – data (see Sect. 1.1). The mathematical model of a real situation needs, as a rule, much more idealization and neglections as can be "compensated" by an exact specification of data incertainty by intervals.

3.2 Specification by Fuzzy Sets

World War Two gave the development of system and control theory for automation an immense impetus. For military purposes always more and more extensive and complicated problems were treated and partly solved employing and using gigantic masses of manpower, material and money. Nevertheless either the necessary abstractions to specify mathematical models led away farther and farther from the facts of reality or the solutions of the mathematical tasks obtained demanded for the application of complicated approximation procedures, the behaviour of which in concrete situations could not always be assessed. The use of procedures developed in this way during the war for control purposes in simple civilian situations proved rather problematic. Expenditure and benefit were in stark discrepancy, the control procedures were frequently too expensive, sometimes too slow, because of their relative complexity too susceptible to faults and, as a rule, inferior to control by manpower.

In the mid-1960s the time was ripe for a new approach to this problem area. On the one hand the mathematical theory, by the development of multi-valued logic and lattice theory, had come into the position to formulate and treat mathematically the pressing demands of practical application, on the other hand the development of electronical computer technology had made progress to solve mathematically formulated tasks for these new structures within tolerable time. This situation was realized, for the first time, by ZADEH, an American system theoretician, in 1965. Since then his name is closely connected with the development of this field. He put two demanding problems to mathematics and to himself:

The model for the practical situation should be chosen as *a simple as possible* one, it should allow a simple and comprehensible solution of the given problem.

The uncertainty on the practical situation and the data are to be modelled in a manner that a computer can understand *the semantic contents* of the problem and of the question sufficiently well.

This leads to the reasonable maxim:

It should be formulated and specified only so accurately and precisely as it is appropriate to the practical problem with respect to the model as well as to the data. Naturally, in the control problem also the single *control instructions* are data.

ZADEH (see ZADEH (1965)) called his ideas: *Theory of fuzzy sets* being unimpressed by the sometimes pejorative use of the word *fuzzy*. Also today his approach is refused by large sections of mathematicians, sometimes even fighted, but this could neither stop nor slow down its triumphal march in practical applications.

In this section some basic ideas of the theory of fuzzy sets are presented, which may suffice to get an idea of the theory. For further details and more references see BANDEMER/GOTTWALD (1995).

3.2.1 The Idea of a Fuzzy Set

For more than hundred years (CANTOR (1984), (1884)) set theory is established as the general basis of mathematics. Every mathematical statement can be reduced to a statement on the membership of elements from a *universal set*, the set of *all* elements to be considered in a given context in general. In the following the respective appropriate universal set will be called *universe*. The general transformability to statements with respect to set membership is a *theoretically* true statement; an *actual* transformation could be very troublesome or even impracticable.

The membership of an element u of a universe U to a subset A can be indicated by a suitable mark in the "list" of all elements u of U. Without loss of generality this mark can be chosen as the number 1 and for filling up the corresponding column of the list the mark 0 indicates then all those elements u *not* belonging to the set A. The *function* generated in this manner is the well-known *characteristic function* usually denoted by χ:

$$\chi_A(u) = \begin{cases} 1 \text{ , if } u \in A, \\ 0 \text{ , if } u \notin A . \end{cases} \tag{3.11}$$

The *basic* idea is now the allowence for an element u to belong to a set A also *gradually*.

From everyday life it is a familiar experience that in many respects there is no either – or, but there are *degrees* or *grades* and *nuances*. Already the pairs of notions

healthy – ill,
sober – drunk,
learned – uneducated,
clever – stupid

make clear this dilemma of allocation to a set in practical cases. Even if an objective scale does exist, as with time and temperature, similar problems can occur:

A German citizen becomes *of age* at his eighteenth birthday 0 o'clock in the morning, that is determined by a law uniquely and precisely in the interest of legal certainty.

But when becomes a certain in Germany resident human being *full grown* (in the biological sense)?

A thermometer located somewhere in a certain room shows 22 degrees Celsius. The precision for reading it is one half degree.

How many degrees Celsius should be recorded as the temperature of the *whole room*, in which, as is well known, a spatially and timely varying temperature field is established, and how should this record be justified?

Problems of this kind can be treated by specifying of *fuzzy sets*, by which every possible exact value is *evaluated* by a number between 0 and 1, the *degree of acceptance*, as it is called, measuring the acceptance that even this value can be considered as a *representative* in the given statement.

Normally a certain man at the age of forty is surely accepted as "biologically" *full-grown*, whereas an eight year old child surely is *not at all*. Between these two time marks, however, there will be an increasing (at least not decreasing) acceptance for a human being biologically full-grown. For the moment, the problem is (yet) *not* the specification of the degrees of acceptance for the age values between, but the impossibility, like with the becoming of age, to fix a *precise* time, at which the change occurs from the state of being *not at all* full-grown to that of being *totally* full-grown.

With a similar problem one is faced, when a precise temperature is to be specified for a whole room. One has to evaluate the different values from a certain interval of possible temperature by acceptance degrees, with which they are accepted as representatives for the temperature of the room. For this evaluation the temperature read from the thermometer will help only as a point of reference, the precision of this reading will play an only minor role when compared with the possible variation within the room.

Here one can see an interesting connection with the concept of interval specification of imprecision as presented in Subsect. 3.1.2. The interval of possible values there is now endowed with an *evaluating function*, which reflects the degrees of acceptance, the *membership function* μ_A, as it is called, by which a *fuzzy set* $\mathcal{A}$ is defined. The peculiarity of the set $\mathcal{A}$ as a *fuzzy* set will *not* be emphasised additionally by using the same type also in the index of the membership function, for the sake of simplicity of presentation.

A *fuzzy set* $\mathcal{A}$ is hence characterized by its membership function μ_A

$$\mu_A | U \rightarrow [0,1] . \tag{3.12}$$

A usual (*crisp*) set is then a special fuzzy set and the characteristic function χ_A is a special membership function. There is a further close connection with the concept of interval mathematics, even with its generalization to arbitrary including sets. Namely, starting from a fuzzy set $\mathcal{A}$ *crisp* sets A_α are defined, over which the membership function μ_A assumes at least the value $\alpha(\alpha \in [0,1])$, these sets are called α-*cuts* of $\mathcal{A}$. In this manner an equivalent representation of the fuzzy set $\mathcal{A}$ is obtained by the infinite set of crisp sets $\{A_\alpha\}, \alpha \in [0,1]$ lying into one another with decreasing α. Hence for every fixed α a basis is established for a treatment by crisp sets, in the one-dimensional case by intervals. The *equivalence* with the representation by a membership function results from the *inversion formula*

$$\mu_A(u) = \sup_{\alpha \in (0,1]} \alpha \cdot \chi_{A_\alpha}(u) . \tag{3.13}$$

Further inversion formulae can be found in the literature (e.g. BANDE-MER/GOTTWALD (1995)).

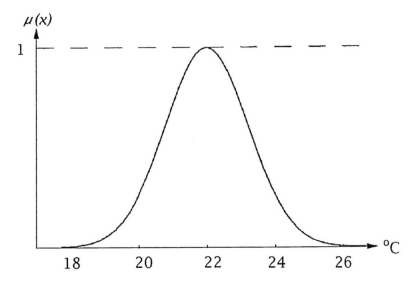

Fig. 3.1. Example of a membership function specified for the fuzzy set "pleasant temperature at a desk in a room"

In appropriate cases this connection makes it possible to choose a finite (usually small) number of α−cuts, for which the given task is treated, and to put together an entire solution by means of the involution formula. But there are also enough procedures, for which such a cutting up is not necessary (see, e.g., Subsect. 3.2.3).

From the concept of fuzzy sets result two basic problems:

In a concrete case for every $u \in U$ a value $\mu_A(u)$ is to be specified, which should express the degree of membership of u to $\mathcal{A}$. This *specification problem* is of central importance for all applications. The next subsection is devoted to this problem.

Since the basic notion of mathematics, *the set*, has got a new definition, the *whole mathematics* must be reconsidered with respect to the consequences, if the usual notion of a set is replaced by the new notion of a fuzzy set. This means, in principle, a *new version* of all notions used in mathematics. This *structure problem* will be a main task for mathematicians, the results of their efforts, developing new procedures and investigating their properties, are of central importance for theory as well as for application. The simplest composition rules and their properties are treated in Subsect. 3.2.3.

3.2.2 Specification of Fuzzy Sets

In the assessment of interval mathematics the difficulty was emphasised to determine *exact* boundaries of those domains, in which the possible values of the considered variables can still vary within the framework of impreciseness.

The problem consists in the fact that the transition from an *absolutely possible* value to an *absolutely impossible* value is effected as an *abrupt* one, such that values of both categories can be infinitely near neighbours, which is scarcely plausible in practical cases.

The possibility to specify the imprecision of an observation or of a measurement by a *fuzzy set* $\mathcal{A}$ removes this dilemma in its *essential* contents. The specification of the membership function μ_A can naturally also *not* be effected without any arbitrariness. But now the possibilities of differently large deviations are weighted and one can specify a *gentle* (gradual) transition. The specification of μ_A should be carried out by a specialist, who is familiar with the given problem. In this manner his knowledge on the situation and on the peculiarities of the measurement or observation process are included, possibly only unconsciously.

On the other hand a specification exactly and uniquely obtained from the given problem, as again and again reminded by theoretical mathematicians, is neither *practically possible* nor *practically useful*. The question here is the specification, the mathematically modelling, but really only the *evaluation* (a mathematical form of assessment) of *impreciseness, uncertainty and vagueness* in a given concrete problem. In this case demanding an exact and unique specification from the situation can be understood only as an unworldly suggestion, especially in view of the usual performance in specifying mixed boundary problems, problems in mathematically programming or a distribution function for a problem from practice in stochastics. In these cases for the sake of tractability neglections and idealizations are so common that they are, as a rule, no longer felt as such ones.

Hence it will be sufficient, if all the persons involved in the project agree on a rough form of the membership functions. This group of people includes persons, who put the problem, as well as those, who treat the tasks, and those, who use the results, and hence all, who have to specify the fuzzy sets from the situation and to interprete and to use the statements obtained by the treatment with fuzzy sets in the given situation. First of all a *local monoticity* of the membership functions will be of importance: The group must agree, in which subdomains the values of a certain membership function should be larger than in other subdomains and whether the membership values should decrease or increase in certain directions. For a mathematical representation *simple* types of functions should be chosen, which can fulfil these considerations sufficiently well. Hence scruples and over-subtlety are not useful in specifying membership functions. The individual "rules" in this *specification* will be reflected in the mathematical form of the *results* and then serve for an adequate *interpretation* and *use* of these "fuzzy" results according to analogous "rules"; hence they remain an *inside* part of the mathematical treatment. Their influence on practical consequences can be neglected, if the fuzzy treatment is carried out in a well-informed manner.

For typical mathematical (crisp) subjects now the fuzzy generalization will be considered.

For the sake of simplicity of presentation and specification the *support* of a fuzzy set A is introduced. The support is a subdomain of the universe, in which the membership to A is *positive*, it is denoted by its abbreviation supp(A)

$$\text{supp}(A) = \{u \in U : \mu_A(u) > 0\} . \tag{3.14}$$

Hence the support includes all elements, which are of interest for the set A and for which the membership is to be really specified.

On the other hand the subdomain of elements is important, which belong to A *completely*, i.e. for which $\mu_A(u) = 1$. This set is called the *core*: core(A)

$$\text{core}(A) = \{u \in U : \mu_A(u) = 1\} . \tag{3.15}$$

A *fuzzy number* is usually defined over the real axis as its universe. If only numbers of elements occur in the set of possible numbers (of persons, objects, etc.) then obviously the set of natural numbers suffice as the universe.

A *fuzzy* number should mathematically specify the idea of "approximately". Hence the corresponding fuzzy set should contain *one and only one* number in its core (e.g. approximately 10). Moreover the support of the fuzzy number is to be specified, i.e. the set of all numbers, which should be accepted as the number in the core to any positive degree. This specification is closely related to the concrete problem considered and must start from it. It is reasonable that the degree of acceptance of the numbers in the support (monotonously) decreases with increasing distance from the number in the core. Also the speed of the decrease depends on the concrete situation. As a rule, a suitable type of functions is chosen describing the speed roughly and containing still some free parameter for an individual adaption. For the fuzzy number "A = approximately 10" over the real axis e.g. the following membership functions are possible

$$\mu_A(x) = \exp\left\{-\frac{(x-10)^2}{a}\right\}; \quad a > 0 \tag{3.16}$$

or

$$\mu_A(x) = \max\{0, \ 1 - b(x-10)^2\}; \quad b > 0 . \tag{3.17}$$

The example shows that rather different types of functions can lead to practically hardly distinguishable specifications by a suitable choice of the parameters. This allows in many cases, for different fuzzy sets occuring in the same problem, to choose the same type of function, this can simplify the practical computations essentially. This is used especially in calculations with fuzzy numbers (see Subsect. 3.2.4).

If the fuzzy number is to be specified only over the set of all natural numbers, then only for (a few) interesting natural numbers of the support the

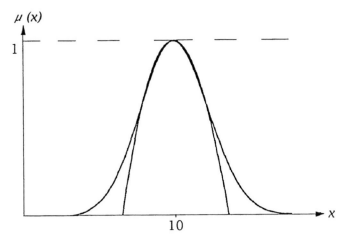

Fig. 3.2. Representation of two membership functions according to (3.16) and (3.17), respectively, with parameters suitable for comparison

membership values are to be determined; obviously it is still $\mu_A(10) = 1$, and the membership must again decrease (monotonously) towards both sides.

In an analogous manner a *fuzzy interval* is specified by fixing a (crisp) interval as the core, from which the membership is decreasing towards both sides (monotonously).

Also the analogous specification of a *fuzzy point* is reasonable. Starting from a crisp point as the core the membership function must decrease (monotonously) towards every direction. Here geometrical domains present themselves as graphs of membership functions, e.g. pyramids and ellipsoids. For construction of *fuzzy vectors*, frequently used in practice, it is necessary to introduce rules for the composition of fuzzy sets, which are presented in the next subsection.

Quite in analogy a *fuzzy domain* is specified by choosing a domain in the usual sense as the core, from which the membership function (monotonously) decreases towards every direction.

For two-dimensional fuzzy points there is a useful visualization. If the membership values are interpreted as *grey tones*, the value 1 as the "darkest" black and the value 0 as the "brightest" white, then a fuzzy point or a (small) fuzzy domain occurs as a greytone spot. This makes it possible not only to get a vivid impression of the "fuzziness" of some specification but also, the other way round, to interpret and treat grey tone pictures as fuzzy sets in suitable situations, a possibility used later on.

But not only mathematical objects can be "fuzzified". It is also possible to specify *verbal statements* by fuzzy sets. This is even emphasised as a special achievement of the new approach, which goes essentially far beyond the possibilities of a "generalized" interval mathematics.

Not all applied branches of science use *quantitatively* mathematically oriented descriptions of processes and situations considered and treated by
them, e.g. in medicine and psychology. But also in everyday industrial work,
processes and courses are described *verbally*, e.g. to develop instructions for
machine operators. The presentation is, according to the target group, written
in a colloquial or a technical terminology, which can be comprehended by the
recipients in its semantic contents. As typical examples one may remember the
instructions for driving a car he got at driving school, especially e.g. the instructions governing the process of parking a car into a (small) parking space
parallel to the traffic direction. Other typical examples are cooking recipes
with their vague data of "a teaspoon of . . . ", others are some characteristic
effects for diagnosing some specific kind of illness, and evaluation of the quality of scientific work or of welded joints or hints for an energy-saving driving
of trains.

In all such cases the essential data as instructions, features, characteristic
values, etc., are given *fuzzily*. The theory of fuzzy sets makes it possible to
bring those *qualitative* descriptions to a mathematical modelling that they
can be "treated" by a computer according to their semantic contents. For this
purpose, in each single case, obviously one has to determine for each occuring
variable (i.e. a quantity, which can assume, in every given situation, a value
from a given set of possible values), which can be its naturally *fuzzy* values and
from which basic domain, the appropriate universe, these fuzzy subdomains
should be taken.

Moreover, such *fuzzy* values of a *fuzzy variable* must be *named* to become
utilizable. But the difference with real numbers is that standard names for
fuzzy sets do not exist (as they do for numbers). From the relation to the
qualitative description of processes and situations is seems reasonable to name
the values of such fuzzy variables immediately by colloquially or technically
usual *words*, where one has to fix, *which* words are allowed to name the fuzzy
values and *which* fuzzy sets should form the range of such a fuzzy variable.

The decision of naming the values of fuzzy variables using suitable words in
our everyday language was already explained by ZADEH (1975), who created
this idea of fuzzy variables, to call them *linguistic variables*, e.g. to distinguish
them from numerical variables with imprecisely determined values. But one
has to be very careful here: it is inessential for the true character of a fuzzy
variable, whether their values are named by linguistic terms or not, as it is
inessential for a real variable to have their values denoted using a decimal or
a dual system of numbers.

The concept of *linguistic variables* gains its attrativeness from the fact that
it allows to "translate" nearly immediately some rough description of some
industrial process (or another situation) – of course a description of such a
process which can successfully be used by a human operator – into a formal-
ized, non-traditional mathematical model to be implemented almost directly.
The *main problem* which remains is to specify the membership functions of the
"linguistic" values which are involved in the modelling process. A theoretical

solution of this problem does not (yet) exist, and it can be expected not in near future, if at all. It seems to be one of the main lines of present research to try to determine these membership functions by automatic learning processes, or even to learn control rules, e.g. by the use of neural net techniques, or by using genetic algorithms (for literature see e.g. BANDEMER/GOTTWALD (1995)). But also here the comment given for the specification of fuzzy numbers and points remains valid that a meticulous consideration of this problem is senseless and of very little use.

The example considered now may be called already "classical", it is the linguistic variable AGE describing the age of some subject or object. It plays its role not only in medical or psychological situations, but also to characterize states of wearing out of tools, e.g. in automated production lines. The values may be fuzzy subsets of the real interval $[0, 100]$, perhaps after an appropriate changing of the scale. Then one could start, e.g. with three different values named *young, middle-aged, old.* The determination of the corresponding membership functions μ_{young}, $\mu_{middle-aged}$, μ_{old} depends essentially on the context. One may compare his ideas of the age of an "old competitive sportsman" with that of an "old cardinal" or even with that of an "old car". The proceeding is analogous to that mentioned with fuzzy numbers: first one has to fix, which values of the scale should be taken at all for each single special value of the linguistic variable (i.e. the supports) and then one has to determine the core of each fuzzy set, i.e. that set of values, which one wants to have in the set as "full members" with firm conviction and absolutely.

If the number of possible values seems to be too poor, the splitting up too coarse, then the theory of fuzzy sets offers the possibility to generate rules to create a "language" (a generative grammar) of fuzzy sets, by which formulations in a colloquial or technical terminology can be translated into such a language of fuzzy sets. Such rules are either *linguistic modifiers* introducing modifications like "very", "more or less", etc., changing the membership function of the linguistic value to be modified systemetically, or *linguistic combination rules* "not", "and", "or", which are carried out by complementation, intersection and union of the corresponding fuzzy sets, or, finally, *rules for quantification and for qualification* of the single fuzzy values: "always", "frequently", "seldom", "never", or, respectively "true", "not very true", "rather false", "absolutely false", or, respectively, "possible", "rather possible", "hardly possible", "impossible".

As an example the determination of values of the linguistic variable SOLUBILITY IN WATER is mentioned, as it was used in a special problem of chemistry.

Finally a case is mentioned, where even physical measurements supply the membership functions of the corresponding values of a linguistic variable. In a chemometric context for the linguistic variable COLOUR in its verbally given nuances *bright yellow, yellow, dark yellow, orange, brick-red, crimson,* etc, as occuring with special chemical indicators the corresponding membership

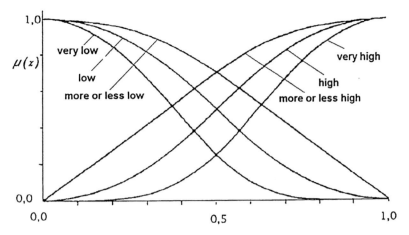

Fig. 3.3. Membership function for the linguistic variable: SOLUBILITY IN WATER (see OTTO/BANDEMER (1988a)), where $x = 0$ means: "insoluble at all", and $x = 1$ means "completely and immediately soluble" on conditions each time to be indicated

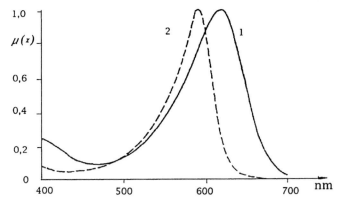

Fig. 3.4. Visible spectra of indicators in the wavelength range between 400 and 700 nm renormed to the height 1 and used as membership functions for two values of the linguistic variable COLOUR: 1 – Bromoscresol (green); 2 – Bromophenol (blue) (with respect to the problem see OTTO/BANDEMER (1988a))

functions were specified as the intensities over the wavelength normed to height 1. Figure 3.4 shows two of the examples.

3.2.3 Operations with Fuzzy Sets

As already mentioned in the preceding subsection the notion "set" forms the basis of mathematics as a whole (CANTOR (1984), (1884)). If this notion is generalized and newly defined, then all operations with sets are to be reconsidered and also newly defined. Since the common (crisp) sets are special fuzzy sets, these new definitions must lead to usual results when applied to crisp sets.

The most elementary operations for usual sets are the union and the intersection of two sets and the complement of a set (i.e. the set of elements of the universe not belonging to the set). The procedures should be carried out elementwise, i.e. the membership of an element of the union, of the intersection or of the complement should depend *only* on its membership in the starting sets. If one puts further reasonable demands with respect to the properties then one obtains the *following suggestions*, which had been presented already by ZADEH in his paper (1965).

For the *union* $\mathcal{A} \cup \mathcal{B}$ of the fuzzy sets $\mathcal{A}, \mathcal{B}$ this is the definition

$$\mathcal{C} = \mathcal{A} \cup \mathcal{B} : \mu_C(x) = \max\{\mu_A(x), \mu_B(x)\} \quad \text{for all} \quad x \in U , \qquad (3.18)$$

and for the *intersection* $\mathcal{A} \cap \mathcal{B}$

$$\mathcal{D} = \mathcal{A} \cap \mathcal{B} : \mu_D(x) = \min\{\mu_A(x), \mu_B(x)\} \quad \text{for all} \quad x \in U \qquad (3.19)$$

and for the *complement* $\mathcal{A}^c$ of a fuzzy set (with respect to the universe U)

$$\mathcal{K} = \mathcal{A}^c : \mu_K(x) = 1 - \mu_A(x) \quad \text{for all} \quad x \in U . \qquad (3.20)$$

For crisp sets the definitions yield the usual results. When the α-cuts are considered, which are crisp sets, then the usual results are obtained, e.g.

$$(A \cup B)_\alpha = A_\alpha \cup B_\alpha . \qquad (3.21)$$

Without any difficulty one can show for the definitions above the following simple calculating rules:

$$\mathcal{A} \cup \mathcal{B} = \mathcal{B} \cup \mathcal{A} , \qquad (3.22)$$
$$\mathcal{A} \cup (\mathcal{B} \cup \mathcal{C}) = (\mathcal{A} \cup \mathcal{B}) \cup \mathcal{C} , \qquad (3.23)$$
$$\mathcal{A} \cup \mathcal{A} = \mathcal{A} . \qquad (3.24)$$

Analogous rules are valid for intersection.

A connection of union and intersection is given by the *deMorgan* laws:

$$(\mathcal{A} \cap \mathcal{B})^c = \mathcal{A}^c \cup \mathcal{B}^c , \qquad (3.25)$$
$$(\mathcal{A} \cup \mathcal{B})^c = \mathcal{A}^c \cap \mathcal{B}^c . \qquad (3.26)$$

If the *inclusion of a fuzzy set* $\mathcal{A}$ in another fuzzy set $\mathcal{B}$ is defined by (as it is also valid for crisp sets)

$$\mathcal{A} \subseteq \mathcal{B} : \mu_A(x) \leq \mu_B(x) \quad \text{for all} \quad x \in U , \qquad (3.27)$$

then one has

$$\mathcal{A} \subseteq \mathcal{B} \Rightarrow \mathcal{A} \cup \mathcal{C} \subseteq \mathcal{B} \cup \mathcal{C} \,, \tag{3.28}$$

and for intersection there holds an analogous statement.

For connecting of two crisp sets A, B defined over different universes X, Y the *Cartesian product* is known, containing all pairs of elements (x, y) with $x \in A$ and $y \in B$. The corresponding notion for fuzzy sets is that of a *fuzzy Cartesian product*, its membership function is defined as

$$\mathcal{C} = \mathcal{A} \otimes \mathcal{B} : \mu_C(x, y) = \min\{\mu_A(x), \mu_B(y)\} \,. \tag{3.29}$$

Also here a series of simple computing rules can be shown (see, e.g., BAN-DEMER/GOTTWALD (1995)).

Though this generalizations of union and intersection for fuzzy sets are simple and natural, there exist also other possibilities for this purpose. However, in this way, some usual and pleasant properties are lost, but this may occasionally be unimportant for an application. One of these possibilities is the *algebraic sum* usable as a union

$$\mathcal{C} = \mathcal{A} + \mathcal{B} : \mu_C(x) = \mu_A(x) + \mu_B(x) - \mu_A(x) \cdot \mu_B(x) \tag{3.30}$$

and the *algebraic product* usable as an intersection

$$\mathcal{D} = \mathcal{A} \bullet \mathcal{B} : \mu_D(x) = \mu_A(x) \cdot \mu_B(x) \,, \tag{3.31}$$

each defined for all $x \in U$.

It belongs to the astonishing deficiencies of the last mentioned definition that the intersection of a *proper* fuzzy set $\mathcal{A}$ with itself does *not* result in the set itself but in a smaller set contained in $\mathcal{A}$. This can be seen from formula (3.31) when considering that the square of a positive number properly less than 1 (the membership value) is smaller than the number itself.

In the course of the recent decades classes of t-norms, as they are called, and suitable to each single one corresponding t-conorms are suggested and investigated, which differ in their algebraic properties. By those means one can try to model connections for practical situations, which differ factually from the usual forms, e.g. for combining of assessments on the credit-worthiness of bank-customers with respect to different forms of security (surety, assets, personal qualities, attractivity of the project intended). Moreover, there are also *compensatory operations*, the results of which lie *between* interaction and union or, more generally, compensate the membership values of the participating sets in a desirable manner. (see for this topic e.g. ZIMMERMANN (1991)).

3.2.4 Connections Via Functions and of Fuzzy Numbers

Besides the generalization of union and intersection of fuzzy sets for application also connections of other kinds are necessary. Let be given a usual (crisp) function f of (several) variables

$$f \mid X \times Y \to Z \; . \tag{3.32}$$

What should happen, if the values of the variables, say x and y, are given only fuzzily, i.e. as *fuzzy sets* $\mathcal{A}$ over X and $\mathcal{B}$ over Y. This corresponds to the situation that x and y are known or can be given only "approximately". One expects that also a fuzzy value $\mathcal{C}$ over Z can be ascribed to the function f as belonging to $\mathcal{A}$ and $\mathcal{B}$. But, how should this fuzzy value be computed? Theoretical considerations from logic and from the theory of relations lead to the proposal due to ZADEH (1975):

$$\mu_C(z) = \sup_{x,y:z=f(x,y)} \min\{\mu_A(x), \mu_B(y)\} \quad \text{for all} \quad z \in Z \; . \tag{3.33}$$

This is written also $\mathcal{C} = f(\mathcal{A}, \mathcal{B})$ shortly and symbolically. This rule can obviously generalized for more than two arguments and is generally called *extension principle.*

For special cases this principle can be simplified essentially with respect to its numerical form. Especially *fuzzy numbers* are considered now, which occur usually as argument values of functions. As already mentioned, a fuzzy number has *one and only one* value in its core and its membership values decrease starting from this value towards both sides (or remain at least constant). Moreover, this fuzzy set has to be *convex*, i.e. every α−cut consists of only one (connected) interval.

For such fuzzy numbers now arithmetical operations are introduced. If the function f in the extension principle is the *sum* $f(\mathcal{A}, \mathcal{B}) = \mathcal{A} \oplus \mathcal{B} = \mathcal{S}$ then the corresponding membership function is

$$\mu_S(z) = \sup_x \min\{\mu_A(x), \mu_B(z - x)\} \; , \tag{3.34}$$

and for the *difference* $\mathcal{A} \ominus \mathcal{B} = \mathcal{D}$ it holds

$$\mu_D(z) = \sup_x \min\{\mu_A(x), \mu_B(x - z)\} \; , \tag{3.35}$$

which was obtained by rearranging of $z = f(x, y)$ with respect to y and inserting into the extension principle. Likewise for the *product* $\mathcal{A} \odot \mathcal{B} = \mathcal{P}$ the membership function is computed using

$$\mu_P(z) = \sup_{x,y:z=xy} \min\{\mu_A(x), \mu_B(y)\} \; . \tag{3.36}$$

In all the three cases obviously one has to add $z \in Z$. The results are always fuzzy numbers, if $\mathcal{A}$ and $\mathcal{B}$ are fuzzy numbers. These formulae can be generalized also for *fuzzy intervals*, in which a crisp interval forms the core instead of a single crisp value. Also here, fuzzy intervals yield again fuzzy intervals.

Also the *negative* $\mathcal{N}$ of a fuzzy number or interval is defined in this way by

$$\mu_N(x) = \mu_A(-x) \quad \text{for all} \quad x \in X \ . \tag{3.37}$$

But one has to be cautious in defining a quotient of fuzzy numbers or intervals, because a division by 0 must be excluded. Hence one expresses first under the additional assumption $0 \notin \text{supp}(\mathcal{B})$ the *reciprocal* $\mathcal{K} := \mathcal{B}^{-1}$ by

$$\mu_K(z) = \begin{cases} \mu_B(1/z) & \text{for all} \quad z \quad \text{with} \quad 1/z \in \text{supp}(\mathcal{B}) \\ 0 & \text{for all other} \quad z \end{cases} \tag{3.38}$$

Using this and again assuming $0 \notin \text{supp}(\mathcal{B})$ one has the *quotient* $\mathcal{Q} := \mathcal{A} \oslash \mathcal{B}$ as $\mathcal{Q} = \mathcal{A} \odot \mathcal{B}^{-1}$ with the membership function

$$\mu_Q(z) = \sup_{x,y:z=x/y} \min\{\mu_A(x), \mu_B(y)\} \ . \tag{3.39}$$

The computing operations for fuzzy numbers and intervals include the usual ones of *interval arithmetics*. Many known rules for *real numbers* hold true for the operations mentioned above, but *not all* of them do. For addition and multiplication one has commutativity and associativity, i.e. one can put in and outside the brackets and change the order of sequence of the operands. But distributivity does *not* hold true *unrestrictedly*. Nevertheless, e.g. in the case that $0 \notin \text{supp}(\mathcal{A})$ as also $0 < \text{supp}(\mathcal{B} \odot \mathcal{C})$ (i.e. this set lies completely within the positive axis), then it holds as usually

$$\mathcal{A} \odot (\mathcal{B} \oplus \mathcal{C}) = (\mathcal{A} \odot \mathcal{B}) \oplus (\mathcal{A} \odot \mathcal{C}) \ . \tag{3.40}$$

Moreover, *in any case* one has instead of the distribution law only the subdistributivity (an inclusion statement)

$$\mathcal{A} \odot (\mathcal{B} \oplus \mathcal{C}) \subseteq (\mathcal{A} \odot \mathcal{B}) \oplus (\mathcal{A} \odot \mathcal{C}) \ . \tag{3.41}$$

Furthermore, one has to have in mind that $-\mathcal{A}$ added to $\mathcal{A}$ does *not* supply the *crisp* zero; hence the two equations $\mathcal{A} \oplus \mathcal{C} = \mathcal{B}$ and $\mathcal{B} \oplus \mathcal{D} = \mathcal{A}$ are *no longer* equivalent formulae (expected is $\mathcal{C} = -\mathcal{D}$), because it may occur that the one equation can have a solution $\mathcal{C}$, whereas there does not exist any fuzzy number $\mathcal{D}$, which satisfies the other equation (with respect to a numerical example see BANDEMER/GOTTWALD (1995)).

This phenomenon occurs already in interval arithmetics. In principle, the splitting up of the computational tasks for fuzzy numbers via consideration of the corresponding α−cuts allows the use of arithmetics for crisp intervals, but in the following more comfortable procedures for arithmetics for fuzzy numbers are presented.

Since the fixing of membership functions in detail can be effected only with certain arbitrariness, which was explained in Subsect. 3.2.1, *practically* it is hardly a general restriction, if for fuzzy numbers in concrete situations the membership functions are restricted to certain *types of functions*. Especially this means that *auxiliary functions* are chosen, *reference functions* L, R, as they are called, assuming the value 1 for the argument value 0 and their values decrease for positive arguments with increasing values *monotonously*. Then one can introduce, using the parameter m for the core value and two positive scale parameter q, p two partial functions

$$\mu_A^l(x) = L((m - x)/p) \quad \text{for all} \quad x \leq m ; \tag{3.42}$$
$$\mu_A^r(x) = R((x - m)/q) \quad \text{for all} \quad x \geq m ,$$

which put together, as a left and a right branch, form the membership function of the fuzzy number $\mathcal{A}$. With L and R fixed, the fuzzy number can be written *symbolically* as

$$\mathcal{A} = (m; p, q)_{LR} . \tag{3.43}$$

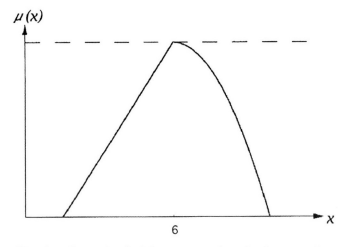

Fig. 3.5. Example of a LR-representation of a fuzzy number

With this symbolism the *addition*, which are represented by the *same* pair of functions LR, i.e. $\mathcal{B} = (n; s, t)_{LR}$, can be written as

$$\mathcal{A} \oplus \mathcal{B} = (m + n; p + s, q + t)_{LR} . \tag{3.44}$$

Also the *multiplication* of a fuzzy number $\mathcal{A}$ by a crisp positive number c can be easily represented as

$$c\mathcal{A} = (cm; cp, cq)_{LR} . \tag{3.45}$$

Because obviously

$$-\mathcal{A} = (-m; q, p)_{RL} \tag{3.46}$$

a symbolically representation of *subtraction* is only possible, if for the number to be subtracted the functions L, R are interchanged. Hence an easy general application can only be expected for $L = R$.

The *product* of two $LR-$numbers is in general *no longer* such a number. A practically usable and easily computable approximating representation in $LR-$form for the product is given by DUBOIS/PRADE (1978), (1980)

$$\mathcal{A} \odot \mathcal{B} \approx (mn; ms + np, mt + nq)_{LR} \qquad (3.47)$$

where especially $0 \notin \operatorname{supp}(\mathcal{A})$ and $0 \notin \operatorname{supp}(\mathcal{B})$ and further $m, n > 0$ and q and t have to be small each, when compared with m and n, respectively. In similar manner approximating formulae are obtained for similar cases.

Starting point for considerations with respect to *division* are again approximating representations for the *inverse*, especially for the case that $0 \notin \operatorname{supp}(\mathcal{B})$ in the neighbourhood of n

$$\mathcal{B}^{-1} \approx (1/n; t/n^2, s/n^2)_{RL} \ . \qquad (3.48)$$

Further approximation formulae can be found in DUBOIS/PRADE (1980) and KAUFMANN/GUPTA (1985). When such formulae are needed, already contained in the software tool or to be introduced by user's programming, one should get a general idea of the numerical precision for the practically required domain, if necessary by a series of suitable examples.

The computations present themselves as particularly easy and clear, if the reference functions are *linear*. Those functions are called *triangular* or simply "hats". They can be characterized again by three numbers, besides the core m the two numbers a_1 and a_2 are given, where each of the straight lines meets the $x-$axis, respectively. Obviously it holds $\operatorname{supp}(\mathcal{A}) = (a_1, a_2)$ and the fuzzy number can be simply represented by

$$\mathcal{A} = \langle m; a_1, a_2 \rangle \ . \qquad (3.49)$$

A connection with the representation according to (3.43) can be seen as follows: With the denotations $L(x) = 1 - bx$ and $R(x) = 1 - cx$ one has $p = b(m - a_1)$ and $q = c(a_2 - m)$.

With $\mathcal{B} = \langle n; b_1, b_2 \rangle$ one can write

$$\mathcal{A} \oplus \mathcal{B} = \langle m + n; a_1 + b_1, a_2 + b_2 \rangle , \qquad (3.50)$$

$$\mathcal{A} \ominus \mathcal{B} = \langle m - n; a_1 - b_2, a_2 - b_1 \rangle , \qquad (3.51)$$

$$-\mathcal{A} = \langle -m; -a_2, -a_1 \rangle \ . \qquad (3.52)$$

As it is expected, however, *product* and *quotient* are *no longer* hat-numbers. Again approximation formulae may help, e.g. for $a_1, a_2 \geq 0$ and with division additionally $0 \notin \operatorname{supp}(\mathcal{B})$

$$\mathcal{A} \odot \mathcal{B} \approx \langle m \cdot n; a_1 \cdot b_1, a_2 \cdot b_2 \rangle , \qquad (3.53)$$

$$\mathcal{A} \oslash \mathcal{B} \approx \langle m/n; a_1/b_2, a_2/b_1 \rangle \ . \qquad (3.54)$$

Also here for further cases it is referred to the software or to the literature mentioned above.

Because of this easy "arithmetics" hat-numbers are very popular in application, although sometimes this means a rather generous idealization. But frequently one can justify this form of membership function from the situation directly (examples see BANDEMER/KRAUT (1990) and BANDEMER/LORENZ (1998)).

3.2.5 Fuzzy Relations

Not only mathematical objects as sets, numbers or points can be "fuzzified" but also relations between objects as known from mathematics. So *equality* between numbers can be represented by the *equality relation*, which is denoted symbolically by the equal sign $=$. The relation is defined over all pairs of numbers (a, b) by declaration for every single pair, whether the relation is fulfilled or not. So the equality relation is *not* fulfilled for the pair $(2, 1)$, whereas it is for $(3, 3)$ obviously. The equality relation $R_=$ can be identified with the set of all pairs of numbers, for which it is fulfilled:

$$R_= : \{(x, y) : x = y\} \ . \tag{3.55}$$

The graphical representation is then the well known bisectrix $x = y$ in the (x, y)-plane. By passing over to the *fuzzy relation* $R_\approx$ "approximately equal" additionally the points in the neighbourhood of this straight line are considered, but with *grading* membership. The determination of this membership by the membership function $\mu_\approx$ should express the idea of "approximately", i.e. it should evaluate the deviation from the *exact* equality with respect to its tolerability in the sense of the desired idea of "approximately". This can, e.g., be effected by determination of an order of fading away of the membership function, which corresponds with the ideas and necessities of the practical problem. The result could be, e.g.,

$$\mu_\approx(x, y) = \left[1 - a|x - y|\right]^+ = \max\left\{0, \ 1 - a|x - y|\right\}, \quad a > 0 \ , \tag{3.56}$$

and represents a linear fading away with a factor a. Also other specifications make sense, which consider the difference in relation with the absolute values, e.g.

$$\mu_\approx(x, y) = 1 - \frac{b(x - y)^2}{(1 + x^2 + y^2)}; \quad b \in (0, 1) \ . \tag{3.57}$$

In analogy with the identification of a crisp relation with the set of all pairs, for which it is fulfilled, one can identify a fuzzy relation with the *fuzzy set*, by which the pairs (x, y) are evaluated with respect to its fulfilment, which is, in the preceding example, specified by the membership function $\mu_\approx$.

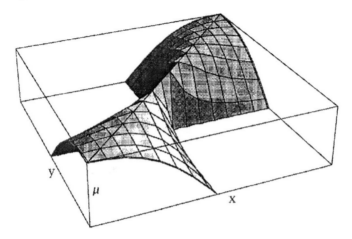

Fig. 3.6. Representation of the membership function for the fuzzy relation "approximately equal" according to (3.57)

According to this model the *crisp* relation $\leq$ ("smaller or equal")

$$R_{\leq} : \{(x,y) : x \leq y\} , \tag{3.58}$$

which corresponds to the halfplane above and on the bisectrix $x = y$, can be replaced by the *fuzzy relation* $\mathcal{R}_{\leq\approx}$ ("more or less smaller than"). The wording "more or less" will be interpreted here in the sense that a sleight surplus is tolerable, hence the halfplane will be endowed with a "fuzziness band" downwards. A specification in analogy with (3.56) would be

$$\mu_{\leq\approx}(x,y) = \begin{cases} [1 - a|x-y|]^{+} & \text{for} \quad x > y \\ 1 & \text{for} \quad x \leq y . \end{cases} \tag{3.59}$$

An other interpretation of the relation can be obtained, if the coordinate axes are understood as domains of two real variable u, v connected by the considered relation. When fixing one of the variables, e.g., $v = y_0$, then the relation $\mathcal{R}_{\leq\approx}$ acts as a *fuzzy bound* $\mathcal{S}_u$ for the other variable, here for u, with the membership function $\mu_u(x) = \mu_{\leq\approx}(x, y_0)$. Such fuzzy analoga of mathematical objects will be used to specify fuzzy versions of mathematical programming topics, where bounds usually play a main role for the solutions (A presentation of fuzzy mathematical programming can be found e.g. in ROMMELFANGER (1994).)

A fuzzy relation shows *to which a degree* two elements of a universe (or a pair each one from a universe) stand in this relation.

With respect to *similarity* of objects such a grading is generally usual: *very similar, not much similar* with the extreme assessments *indistinguishable, totally different*. A relation expressing such a relation is called *similarity relation*. Such a relation $\mathcal{R}$ is expected to be *reflexive* (i.e. every element is indistiguishable from itself: $\mu_R(x, x) = 1$) and *symmetrically* (i.e. every element x is similar to every element y in the same degree as this y is to x: $\mu_R(x, y) = \mu_R(y, x)$).

Within the set of crisp relations especially *equivalence relations* are considered, which split the universe into disjoint subdomains each of all elements, which are equivalent with each other. These relations possess, besides the properties of reflexivity and symmetry, also the property of transitivity: if two elements each are equivalent to a third one, then they are also equivalent with each other. As is well known this does not hold with respect to similarity, similarity can diminish in a series of pairwise comparisons. This may become clear, when imaging a unit of soldiers fallen in line and were compared by individual height of neighbours. Now it is tried to define transitivity for fuzzy sets anew, to come to partitions of the universe as they are known from crisp relations (see for more details e.g. in BANDEMER/GOTTWALD (1995)). Similarity relations will play their important role in fuzzy qualitative analysis in Chap. 6.

For the development of *fuzzy controler* from verbally formulated rule bases, in which these rules are represented by a relational connection $\mathcal{R}$ of linguistic variables (for the fuzzy input $\mathcal{A}_i$ and the fuzzy output $\mathcal{B}_i$ of the controler),

$$\mathcal{B}_i = \mathcal{A}_i \circ \mathcal{R}; \quad i = 1, \ldots, n , \tag{3.60}$$

the problem is to solve this system of relational equations. This is the starting point for that, what is not quite exactly called *fuzzy logic*. Some hints to the extensive literature on fuzzy control are contained e.g. in BANDEMER/GOTTWALD (1995).

4

Specification and Use of Uncertain Variability

In the preceding Chap. 2 important kinds of uncertainty are treated. First the *impreciseness* of mathematical quantities was considered: *it is uncertain, what had been precisely given* (by measurement or observation). This kind of uncertainty was handled by interval mathematics and by specification of fuzzy sets. Then the case was looked at that uncertainty is due to the *verbal* formulation of the findings *by observation and measurement*, i.e. from the *vagueness* of language: *it is uncertain, what is accurately meant*. Also here the specification of fuzzy sets suggests itself, now as specification of values of *linguistic variables*, as they are called.

In both cases the currently given situation was of interest. Statements about *similar* situations in future are only justified by the general scientific mode of conclusion: *Similar results in similar situations*. Now the *uncertainty in the future* should supply the matter of interest: *it is at present uncertain, what will be in future*. This is the province of *stochastics*, the theory of the regularities of variability. The basic notions of this theory are *chance* and *probability*, which should be presented in the following first section.

4.1 Chance and Probability

4.1.1 Model Ideas for the Notion Chance

Unique events in the past or in the future are objects of interpretation and speculation more than of formulation of theories and prognosis. Hence frequent events, occuring "in masses", are suitable for a systematic investigation to find "regularities" and to give some hope for a possibility to predict their behaviour in future.

These regularities do not refer to a *single* concrete situation, but to large numbers of such situations "of the same kind", as it is known from playing at dice: The result of the next throw cannot be predicted, but one expects that a (fair) die yields each single of the six possible outcomes with (approximately) equal frequencies in the long run.

All the notions above, which have been marked with inverted commas, must be stated more precisely *within the framework of mathematics* to get a theory, which base its statements *scientifically*. By this process of "making scientific" the meaning of certain words and wordings from everyday language will *change* their semantic contents when used in the context of this theory and its application. Usually, especially in the routine software, it is *not* referred to this fact explicitly. This may be the source for many and extensive misinterpretations of the procedures and their results, which leads sometimes to senseless applications and absurd conclusions. Hence the philosophical and mathematical basic assumptions and ideas will be discussed relatively broadly at this place.

If events occuring "in masses" should be considered, first one has to take care for the *comparability* of the single events.

What should be taken as *essential* and *fixed*?

For this purpose the notion *random experiment* is introduced, following the terminology of physics and chemistry: A random experiment is defined and given by its *fixed conditions*. (It is even spoken of *experimental conditions*).

These conditions should be fullfiled, i.e. the experiment can be performed, arbitrarily often, (because of the mathematical theory!), hence the experiment can be repeated arbitrarily often.

The repetition of an experiment, a single performance of it, consists in the fulfilment of the fixed conditions and in the recording of the phenomena observed, i.e. the outcome of that trial.

Obviously, to be worth to be observed at all, an experiment must have at least two different possible outcomes. In this case, however, the fixed conditions *cannot* determine the observed outcome *uniquely*!

Now, the *mathematical model idea* consists in that all conditions, which are *not* fixed, but influence the outcome, are explained as *the effect of chance*.

Hence the explanation of the notion *chance* differs essentially from its semantic contents in everyday life. There is *no* flavour of rareness and unexpectedness, but it is a simple expression of ignorance on what will influence and decide on a coming observed outcome of the experiment at last.

A further aspect of the notion "chance" in the framework of the mathematical theory and its applications should be really noticed: By the fixing of the conditions the user *decides*, *what* he will regard as the *effect of chance* in the given situation. This can be either an expression of a *being not able to know* (further conditions are not accessible for a fixing) or of a *being not interested to know* (further conditions should not be fixed, e.g. because of high costs of fixing *or* such conditions should vary "by chance", in order to obtain statements from the mathematical investigation being more generally valid).

This formation of the concept is of central importance for all applications and plays a main role in the capture of meaningful data for mathematical investigations.

In a next step it is to be fixed, what should be considered as a *result* of a certain *experiment*. All single possible outcomes of trials of the random experiment and all single sets of different outcomes are called *random events* and are denoted for distinction by different characters $A, B, C, \ldots$. The set of *all possible* events of a random experiment united forms the *certain event* Ω, as it is called, the *empty* set of events is the *impossible event* $\emptyset$, as it is called. This concept makes it possible to represent mathematically interesting situations between events.

All random events and the both extreme events mentioned above form the field of events $I\!A$ of the experiment, which so contains all what is of interest with respect to the experiment.

This field of events is interpreted, as a rule, as a system of sets, in which Ω represents the universal set and $\emptyset$ the empty set, and the random events consist each of *elementary events* $\omega \in \Omega$ or of coarser disjoint subsets of Ω, which are called *atoms*. (The set Ω corresponds to the *universe* in the definition of fuzzy sets. The denotion by Ω is but traditional in stochastics and hence should be kept in this context to facilitate the access to textbooks.)

The usual operations with sets can now be interpreted in the language of the corresponding random events:

The *intersection* of two events $(A \cap B)$ means that both the events occur *simultaneously*.

The *union* of two events $(A \cup B)$ means that *at least one* of the both events occurs.

The *complement* A^c of A with respect to Ω means that the event A does *not* occur.

Finally, the *inclusion* $(A \subseteq B)$ means that together with the event A also the event B occurs: A is followed by B.

4.1.2 Probability

Seen in the light of history the first determination of future uncertainty of random events considered cases, in which each single of the finite many elementary events has the same chance for its occurrence in a future trial. Instead of elementary events also atoms A_i with the same property were handled, which form a partition of the certain event $(\Omega = \cup_i A_i)$. A normalization to the chance measure 1 for the certain event led then to the well known classical definition of *probability* as a normalized chance measure:

The *probability* $P(B)$ *of an event* B is equal to the *number* of all atoms A_i, to which this event B follows, *divided* by the *number* of *all* atoms, which form the partition of the certain event.

This rule for a determination of a probability does only hold in the case that a *finite* system of disjoint atoms (or elementary events) is given, which form a partition of the certain event and of which each single has the *same* chance to occur. The applications of this rule in the investigation of the game of dice or of Lotto, and in quality control are well known (see some textbook for details).

If one keeps the assumption of equal chances, but allows as elementary events all points of a continuum, e.g. a plane domain, then one obtains in generalization of the classical probability the *geometrical probability*:

The *probability $P(B)$ of an event B* is equal to the *area* of all its elementary events divided by the *area* of Ω.

This is the starting point of *stochastic geometry*, which is devoted to spatial distributions and geometrical shapes of objects both influenced by chance, e.g. in medicine and earth sciences (see, e.g. STOYAN/KENDALL/MECKE (1995)).

In general there is neither an atomic structure nor a principle of equal chances. Hence VON MISES in his paper (1919) suggested for this case to choose as the probability of an event B the limit of its relative frequency $W_n(B)$ after n trials:

$$P(B) = \lim_{n \to \infty} W_n(B) . \tag{4.1}$$

For this purpose some additional theoretical assumptions must be put on the manner, in which the series of trials should be performed and the limit should be interpreted. With these problems some researchers within mathematics are still concerned from time to time, but which might be without any interest in the present context. This discussion on the basis of probability faded out, when KOLMOGOROV in his book (1933) *axiomatized* simply some properties of relative frequency for a definition of probability and introduced it as a normalized measure over the experimental domain. Besides the normalization $P(\Omega) = 1$ it is the addition rule for mutually disjoint events (i.e. which cannot occur simultaneously: $A \cap B = \emptyset$):

$$A \cap B = \emptyset \Rightarrow P(A \cup B) = P(A) + P(B) , \tag{4.2}$$

which governs the so created *probability theory*. From this rule many other computing rules can be derived, of which only the *general addition rule* is mentioned here:

$$P(A \cup B) = P(A) + P(B) - P(A \cap B) . \tag{4.3}$$

A very fruitful concept proved that of *conditional probability*, which asks for the probability of an event B, if another event A has already occurred or can be so conceived

$$P(B|A) = \frac{P(A \cap B)}{P(A)} , \tag{4.4}$$

where, obviously, $P(A) \neq 0$ is to be assumed.

With the definitions up to now the idea plays some role that the *proba-bility* represents a looking ahead on the *expected frequency* of the event in a sufficiently long series of trials, hence it reflects an *objective* situation. This is called the *frequentistic interpretation of probability*.

Already very early (LEIBNIZ (1703)) in the treatment of uncertainty (see for the accurate sources e.g. SCHNEIDER (1989)) the notion probability was used also as a degree of belief, of conjecture, of doubt, of uncertainty, and of suspiciousness. Then, however, probability is a matter of an "individual" that believes, conjectures or takes for possible. This *subjective interpretation* as a measure of the *feeling of uncertainty* has also its followers (among others DEFINETTI (1937), SAVAGE (1972)).

The problem left then is still the *specification*, i.e. the subjective deter-mination of the values $P(A)$ for all interesting events A. For this purpose it was suggested to use the *betting behaviour* of the individual: The numeri-cal value of a probability should be specified as proportional to the sum an individual would be willing to pay should a probable event A, expected by him, *not occur*, i.e. a proposition that he asserts proves false. Naturally, it must be assumed that such a sum can be specified. Then it is shown that the subjective measure of uncertainty so defined obeys the axioms of probability theory, provided that the behaviour of the individual satisfies somes condi-tions of "rationality" (see SAVAGE (1972)), first of all the rule of coherence. On this basis, the subjectivists succeeded in showing that KOLMOGOROV's axioms were the *only* reasonable basis for evaluating subjective uncertainty.

This rather extreme attitude can be contested from a philosophical and from a practical point of view (DUBOIS/PRADE (1988)), from which some remarks may indicate the problem area.

So it seems difficult to maintain that every uncertain judgement obeys the rules of betting. The necessary monetary commitment that forms an essential part of the model could prevent an individual from uncovering the true state of his knowledge, for fear of financial loss.

From the practical point of view, it is clear that the number given by individuals to describe, in terms of probabilities, for example the state of their knowledge, must be considered for what they are, namely, *approximate* indications. This is admitted even by SAVAGE explicitly, nevertheless it is demanded that a rational individual must be able to furnish precise numbers, when proper procedures for their elicitation are used.

Not willing to enter this controversy, for application it seems advisable in every case to consider the probabilities specified subjectively as (perhaps only rough) approximations, the character of which is transmitted, naturally, to the conclusions drawn from it, e.g. by computing rules for probabilities.

An important strategy to adapt uncertain knowledge or estimation to given experimental findings is the use of BAYES' theorem. The idea standing behind is the assumption that it is possible to specify an "estimation" of the probabil-ity $P(A_i); i = 1, \ldots, N$, for each single event of a partition of the certain event (i.e. a system of mutually disjoint events, $A_1, A_2, \ldots, A_N$). These estimations should reflect the probability or simply only the subjectively specified state of information *before* at present the investigation is to be performed. Hence it is called *a-prior probability*. Then it is conceived that the experiment is performed, to which the events belong, and that the event B is *realized*. This result contains information with respect to the experiment and hence to the events A_i. How should the probabilities of these events be evaluated *after* this result? The answer to this question is given by the BAYES' theorem

$$P(A_i|B) = \frac{P(B|A_i)P(A_i)}{\sum_{j=1}^{N} P(B|A_j)P(A_j)} \ , \qquad (4.5)$$

which defines the *a-posteriori probability* (for details with respect to the derivation see some textbook or ROBERT (2001)).

Because this formula can be applied sequentially, i.e. repeatedly after every result of a further trial, it establishes the basis of a *learning theory*, which is at present practised by the type of the probabilistic *neural networks* (see also Subsect. 6.1.5). Also in *information theory* the formula serves as a basis for decisions. Information theory starts with considering the following situation: Given *outgoing signals* A_i through a *randomly disturbed channel* K, which is characterized by the probability law $P(B_j|A_i)$ (the channel property, as it is called), are changed into *incoming signals* B_j. The *maximum* a-posteriori probability $P(A_i|B_j)$ is then the criterion, which of the outgoing signals A_i should be assumed to belong to the observed incoming signal B_j. Within the BAYESIAN *decision theory* the decisions to be made *after* an observation and using a loss function are evaluated by this a-posteriori probabilities (see for details e.g. BERGER (1985) and Sect. 4.3).

In every case of application it should be really noticed that a-priori probabilities are only rather rough data, a high precision in numerical handling them, e.g. in the BAYES' theorem, does not make sense. One should always check the meaningfulness of the results obtained within the practical context.

From the definition of conditional probability (4.4) another central notion of stochastics is deduced, the *pairwise independence* of two events A and B. The notion reflects the fact that the occurrence of the one event does *not* have any influence on the probability of the other one, that $P(A|B) = P(A)$ as well as $P(B|A) = P(B)$ are valid. Together these two formulae yield the essentially simpler one

$$P(A \cap B) = P(A) \cdot P(B) \ , \qquad (4.6)$$

which is therefore chosen as the *definition* of the pairwise independence of two events. The notion of independence *of more than two* events will play an important role for the probabilistic inference from samples (see Sect. 4.2), especially in application.

4.1.3 Random Variables and Their Distributions

As already mentioned in the preceding subsection random events can be interpreted as *sets*, more precisely as subsets of the certain event Ω. With respect
to a visualization of models as well as for handling it is convenient, if sets on
the real axis can be chosen. For application it is rather unimportant, whether
these sets are understood as direct equivalents or as mappings (from the field
of events $I\!A$). For a theoretical treatment, however, the idea of mapping is
useful. Mostly with the representation of the events as sets on the real axis an
idea of the *genesis* of randomness is connected, as will be shown by a simple
example.

An experiment is considered, in which only one single random event A
is of interest. The field of events $I\!A$ has then the form $\{\emptyset, A, A^c, \Omega\}$. This
experiment is repeated n-times *independently*. For this purpose the notion of
pairwise independence of n events in the sense of (4.6) must be *generalized* to
the notion of *independence* for all n events: n events from n fields of events are
independent, if for any selection of one event from each of m of these fields
($m = 2, 3, \ldots, n$) the probability value for the *simultaneous* occurrence of
these selected events is equal to the product of their single probability values
(see details in some textbook under the headword: independence of events).
Note that pairwise independence of three or more events does not imply their
independence in this sense.

Given the n results of the trials, it is counted, *how often* the event A has
occurred. This number X is *at random*, since it depends on the random results
in the single trials. Hence it is called a (one-dimensional) *random variable*.
It can, obviously, assume *each* of the values $0, 1, \ldots, n$ (its domain), naturally
randomly for every series of n trials according to the experiment.

Let be the probability of A equal to p in one trial. Then by combinatorical
considerations and using the assumption of independence one can compute
the probability values for each single value of X, because $X = i; i = 1, \ldots, n$,
are all random events:

$$P(X = i) = \binom{n}{i} p^i (1 - p)^{n-i} . \tag{4.7}$$

The random variable X can assume only the $n+1$ values $i = 0, \ldots, n$. Two
or more of these events $X = i$ are simultaneously impossible, hence they are
incompatible or disjoint. Their union is the certain event and hence the sum of
their probability values is 1. The probability 1 of the certain event is *distributed*
among the single events $X = i$. Therefore the assignment of probability to the
values of a random variable is called a *probability distribution*, or shortly a
distribution.

If, as in the preceding example, the random variable X can assume only
finite many discrete values, then it is called *discrete random variable* and its
distribution a *discrete distribution*. This name remains obviously meaningful,

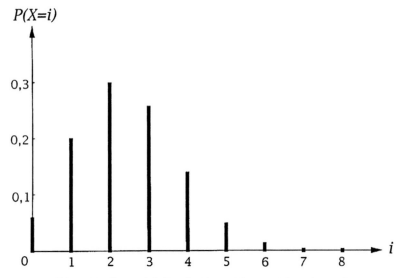

Fig. 4.1. Binomial distribution with $p = 0,3$ and $n = 8$

if the random variable can assume denumerably infinite many values, e.g. all natural numbers.

The distribution types bear mostly more or less easily remembered names. So, the example distribution above is called *binomial distribution*, because of the binomial coefficients in its formula. Naturally, the number values of the probabilities in (4.7) are also dependent on the probability p of the basing event A and the number n of the trials. Hence p and n are called the *parameters* of the distribution. It is liked to represent the probabilities for given parameter values graphically, as is shown e.g. in Fig. 4.1. (In some textbooks the tops of the value columns are connected by segments, but this is absurd, because probabilities for values *between* the natural numbers are obviously senseless.)

Besides the binomial distribution, which is used especially in statistical quality control, there are also several other types of discrete distributions, which are of interest for application. One of them is the *hypergeometrical distribution*, which yields e.g. the probabilities for winning combinations in Lotto; another one is the *Poisson distribution*, a limiting distribution of the binomial distribution, if n tends towards infinity and simultaneously p tends towards zero, which plays a role especially for events with very small probabilities in but numerous trials:

$$P(X = i) = \frac{\lambda^i}{i!} e^{-\lambda} \quad \text{mit} \quad i = 0, 1, 2, \dots \tag{4.8}$$

Provinces for application of Poisson distributions are the atomic decay, then also queueing processes with high numbers of clientes and rare demands, and particle structures in materials in engineering and medicine.

With respect to further suggestions for distributions in textbooks and in software one should be concerned first of all with the model ideas on the genesis of the corresponding random variables in assumed "standard situations", which stand behind the formulae for the given distributions. These ideas contain essential hints for the usability of the distributions in practical situations and on the "experimental conditions" to be observed.

Naturally, in *every* practical situation there are each only *finite many* distinguishable results. In principle, it would be sufficient to consider only *discrete* random variables. This can, however, lead to rather confused and extensive representations, particularly with measure and observation processes with relatively and perhaps necessarily high precision. Moreover, perhaps available knowledge on the processes, and hence on the random genesis of the measure and observation values, *cannot* be introduced into the model. Therefore, in these cases it is supposed that the considered random variable can assume values from a *continuum*. This can frequently be justified also objectively, if the values express quantities of time, length, temperature, pressure, or mass. But even with sufficiently high amounts of money the assumption of continuity on the cent level seems meaningful.

Random variables with such continuous domains are called *continuous random variables*. The probability of a single value will then vanish, as a rule, without that the value becomes impossible. Note that actually the probability of the impossible event is zero, but that with a continuous domain an event with vanishing probability needs *not* be impossible. Because of the uncertainty (fuzziness) of observations mentioned at the beginning and considered already earlier (see Sect. 3.1) only intervals are of interest anyway. Instead of events $X = i$ now events of the form $X \in I$ (the random variable X assumes a value from the interval I), are those, the probability of which is to be determined.

For this purpose the existence of a function f is supposed, which measures the "intensity" of the probability, such that the "infinitesimal" interval $[x, x + dx]$ is endowed with the infinitesimal probability $f(x)dx$. For a finite interval $I = [a, b]$ the probability

$$P(a \leq X \leq b) = \int_a^b f(x)\mathrm{d}x \qquad (4.9)$$

is obtained; even if one or both of the ends of the interval tends towards infinity. Obviously, the function f, called the *probability density* or shortly *density*, must have some natural properties: it must not be negative, it must be integrable and its integral over the whole axis, corresponding to the certain event, must have the value 1. Since the probability of a single value for continuous random variables is always zero, it remains unimportant, whether for the event in brackets in (4.9) one or two of the endpoints of the interval are included or not.

The best known example of a continuous distribution is the suggestion of GAUSS: the normal distribution, as it is called,

$$f(x) = \frac{1}{\sigma\sqrt{2\pi}} \exp\left\{ -\frac{(x-\mu)^2}{2\sigma^2} \right\} \tag{4.10}$$

with its parameters μ and σ (an example see in Fig. 4.2).

It came out of a limiting consideration, which will be treated in the next subsection.

For a long time it was presumed that *all* randomly influenced measure and observation processes show a probability distribution of this type. This is really not so, but frequently the normal distribution remains a useful approximation. One can easily realize that the normal distribution is always an *approximation*. This is not only because of the always existing observation uncertainty (fuzziness), but also because it assigns *positive* probabilities to intervals *everywhere* on the axis, e.g. also to *negative* values of mass and length. Certainly, in practical applications this is rarely of importance, since the distribution concentrates the probability amount 0.9973 to the interval $[\mu - 3\sigma, \mu + 3\sigma]$. Usually the parameter μ is called the *"true" measurement* and the parameter σ a *measure of precision of the measurement*.

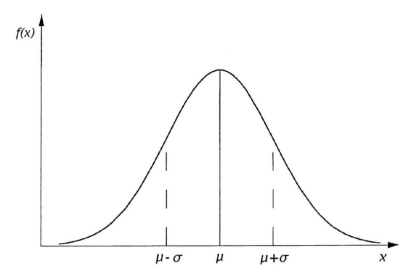

Fig. 4.2. Density of a normal distribution

A pleasant property of the normal distribution is that after a changing of the origin and of the scale the random variable remains normally distributed. Hence the random variable X can be standardized by the linear transformation

$$Y = \frac{X - \mu}{\sigma}. \tag{4.11}$$

The random variable Y has then the parameter values $\mu = 0, \sigma = 1$.

Having realized that the normal distribution does not describe *all* randomly influenced measurements and observations, one turned to functions of random variables hoping to obtain for these functions better approximations by normal distributions. From the results of this way the logarithmic normal distribution proved very useful.

Besides, mainly from interesting application cases, also further types of continuous distributions were established, with respect to which it is referred to the textbooks. Before using them one should first of all look at the model ideas from which they were derived. As an example merely the *exponential distribution* will be mentioned here, which is defind by the simple density

$$f(x) = \lambda \exp(-\lambda x) , \tag{4.12}$$

with the parameter λ and the domain $(0, \infty)$. It has a remarkable property. If one takes this distribution as a model for a random length of time T and asks for the conditional probability that this length after a duration of t units will last still another s units, i.e.

$$P(T \geq t + s | T \geq t) = \frac{P(T \geq t + s)}{P(T \geq t)} , \tag{4.13}$$

then one obtains, by inserting the corresponding integrals over the simple density (4.12), that this probability equals the probability $P(T \geq s)$, as if the length of time has started in the very moment t. Hence the distribution is called "without memory", it does not remember the length of time already passed. According to extensive practical investigations the length of telephone conversations obeys in each case sufficiently precisely an exponential distribution. Therefore the expected length of such a call could not be estimated more precisely if one would know how long the conversation has lasted already, an interesting question for a waiting customer.

Moreover, there is a useful connection between a Poisson distributed random variable and an exponential distributed one. If one considers "point shaped" events (e.g. times of arrival) on the time axis and realizes that their number in periods of fixed length obeys a Poisson distribution, then the distances of such (following one another) point shaped events is exponentially distributed. Therefore, the outlook on the problem can be chosen from each side without any loss of information.

All the mentioned distributions in closed analytical form were derived each one on special assumptions, which will be satisfied only approximately in practical cases. Hence it does not make sense to handle probabilities with high precision in such environments. When such items are found in concrete problems, one should question the precision and round the values reasonably.

For a unified representation of probability distributions for discrete as well as for continuous random variables the *distribution function* $F_X(x)$ was introduced, which indicates the probability of the random event $-\infty < X < x$. For *continuous* random variables this is the integral

$$F_X(x) = \int_{-\infty}^{x} f(x)\mathrm{d}x \tag{4.14}$$

and for *discrete* random variables the corresponding sum

$$F_X(x) = \sum_{a_i < x} P(X = a_i) , \tag{4.15}$$

where the a_i run through the domain of X.

With some other mathematical notions the representation can be unified further and even mixed distributions (i.e. with a discrete part and a continuous one, e.g. lifetime distributions with a stillbirth portion) can be specified in closed form. The distribution function is obviously monotonically nondecreasing (from 0 to 1) and continuous on the left, i.e. in a jump the value of the function lies at the *lower* border. For discrete distribution the distribution function is a *step function* with the jumps at a_i of height p_i: $p_i = P(X = a_i)$. For further details it is referred to the textbooks.

In long series of trials the arithmetic mean of the results is an interesting quantity. If it occurs as the average payoff in a (randomly influenced) zero-sum two-player game, it can be regarded as (the long-term mean) expected gain and can be compared with the stake in every game. If both this values are equal, then the game would be called *fair*, in the other case the game makes a profit (the difference between the expected gain and the stake in every game) on average for one of the two players.

In the next subsection also the behaviour of the relative frequency in series of infinite many trials is considered, which was used already by VON MISES to define probability. In this way it can be shown that the arithmetic mean tends towards a "distribution mean", which is used to define an *expected value* for the distribution. For continuous distributions one obtains

$$\mathrm{E}\,X = \int_{-\infty}^{\infty} x f(x)\mathrm{d}x \tag{4.16}$$

and for discete distributions correspondingly

$$\mathrm{E}\,X = \sum_{i} a_i p_i , \tag{4.17}$$

where the p_i are the probability values for a_i, respectively.

Naturally, it must be assumed that the integral and the sum (for countably infinite many values) yield finite values at all.

In corresponding manner also expected values for functions g(X) of a random variable can be computed, when x and a_i in the formulae are replaced by the function values $g(x)$ and $g(a_i)$, respectively.

As an important special case the function

$$g(x) = (x - \mathrm{E}\,X)^2 \tag{4.18}$$

is considered, which evaluates the square deviation of the random variable X from its expected value $E\,X$. Note that $E\,X$ is not a random variable but a *constant* dependent on X, a characteristic of the random variable X. The expected value of $g(x)$ according to (4.18)

$$E\,g(X) = E\,(X - E\,X)^2 = D^2X \qquad (4.19)$$

is called the *variance* of X and is a measure of variability of the random variable X and of the variations of its realizations (i.e. the results of the trials according to the random experiment, which is represented by the random variable).

As important for application the *rule for the addition of variances* is mentioned. When the sum of random variables, in the simplest case $Z = X+Y$, is considered and the variance of Z is to be computed, then because of the linearity of sums and integrals one obtains for D^2Z

$$D^2Z = D^2X + D^2Y + 2E\Big[(X - E\,X)(Y - E\,Y)\Big]. \qquad (4.20)$$

If X and Y are *independent* of each other or at least are uncorrelated, then the third term on the right side equals to zero and the variances are added. In this context uncorrelatedness means less than independence, namely, only that this third term in the sum disappears for even these two variables X and Y. This case can happen even if the two variables are *not* independent in the sense of probability theory. This argumentation remains valid also for the *difference* $X - Y$, because the subtraction sign occurs then only before that third disappearing term.

This fact is used in the *error propagation law*, as it is called, to evaluate and to assess the influence of small errors in the arguments of a function.

Starting point is the linear part of a TAYLOR's expansion in several variables, e.g.

$$df = \frac{\partial f}{\partial x}dx + \frac{\partial f}{\partial y}dy + \frac{\partial f}{\partial z}dz + \cdots \qquad (4.21)$$

$$= a\,dx + b\,dy + c\,dz + \cdots \qquad (4.22)$$

If the differentials are replaced by corresponding random variables F, X, Y, Z, ..., which should represent the random finite deviations from the "true" function or argument values f, x, y, z, respectively, and if it is assumed that the derivations X, Y, Z, ... are pairwise independent of each other, or at least uncorrelated in the sense given above, then one obtains for their variances

$$D^2F = a^2D^2X + b^2D^2Y + c^2D^2Z + \cdots \qquad (4.23)$$

The variance of the arguments are estimated from the experimental findings (see Subsect. 4.2.2), e.g. by the corresponding sample dispersions $s_X^2, s_Y^2, s_Z^2, \ldots$, and then they yield for the deviation s_F from the function value according to

$$s_F = \sqrt{a^2 s_X^2 + b^2 s_Y^2 + c^2 s_Z^2 + \cdots}, \tag{4.24}$$

where for a statement at the measurement point $(x, y, z, \ldots)$ the $a, b, c, \ldots$ are to be replaced by the according values of the partial derivatives $\frac{\partial f}{\partial x}, \frac{\partial f}{\partial y}, \frac{\partial f}{\partial z}, \ldots$. This is the famous *error propagation law*, as it is called, due to GAUSS. In spite of the scientific claim connected with the name of GAUSS this proceeding is in its essence only a useful heuristic.

Multidimensional random variables as e.g. vectors of random variables are considered in Subsect. 4.2.1; with respect to further details it is referred to the textbooks.

4.1.4 Asymptotic Statements

In the beginning of this chapter stochastics was presented also as a mean to investigate regularities in events occurring "in masses". This involves, naturally, also the consideration of infinite *series of random variables* $X_n; n = 1, 2, \ldots$ and their behaviour in the limit $n \to \infty$. The results can then be used for extensive data sets, which may be assumed to be obtained from such a series.

In stochastics different kinds of *convergence* of such series are distinstiguished: *almost certain convergence*, if the convergence has the probability 1 with respect to all elementary events; *convergence in square mean*, if the expected values of the square deviation from a fixed random variable (the limit in the mean) tend towards zero; and finally *convergence in probability*, if the probability of the absolute deviation from the limit tends towards zero for every elementary event and every arbitrarily small deviation. On certain conditions these three notions of convergence are even equivalent. Their relationship among each other and general assumptions as weak as possible for their occurrence and the rate of convergence are a wide field of research for mathematicians concerned with stochastics. For the user, who will be only seldom in the position to test the partly rather complicated mathematical conditions or only to consider them in the light of the given situation, it would be enough to know that many procedures offered in the software are based on such asymptotic investigations and make sense only for sufficiently extensive data material, especially for samples of large size (With respect to the notion "sample" see the next section.).

Some notions and statements used in this field should now be presented. For further details it is referred to some textbook.

A series of random variables $\{X_n\}$ obeys a *law of large numbers*, if the series of differences of their arithmetic means tends towards zero in one of the senses given above

$$\frac{1}{n} \sum_{i=n}^{n} X_i - \frac{1}{n} \sum_{i=1}^{n} E X_i \to 0. \tag{4.25}$$

If, especially, all the X_i have the *same* expected value $E X$, then this means that the arithmetic mean tends towards this expected value.

Moreover, also the relative frequency of an event A can be considered as an arithmetic mean of a random variable X with the value 1 (the event A occurred) and 0 (it does not). Then the law of large numbers means that the relative frequency converges to its probability. Whereas VON MISES used this statement as the *starting point* for his definition of probability, it is here a *conclusion* from KOLMOGOROV's axioms.

Further-reaching statements can be obtained, if a notion *convergence in distribution functions* is introduced (by means of characteristic functions as they are called). In this manner it can be shown that, with appropriate normalization of the partial sums $Z_n = \sum_{i=1}^{n} X_i$, e.g. with

$$Z_n^* = \frac{Z_n - \mathrm{E}\, Z_n}{\sqrt{D^2 Z_n}} \,, \tag{4.26}$$

the distribution of Z_n^*, on certain conditions, tends towards a standardized *normal distribution* ((4.10) with $\mu = 0$ and $\sigma^2 = 1$). Such statements are called *limit theorems* (according to their considered field of validity a *local* or a *global* one). Especially it follows that even a (discrete) binomial distribution for large n can be approximated very well by a (continuous) normal distribution with the parameters $\mu = np$ and $\sigma^2 = np(1 - p)$.

The property of a normal distribution, on certain partly also restricting conditions, to be *limiting distribution* of suitable normalized sums of random variables, is used with high profit in probabilistic inference and emphasises the privileged role of this distribution. The Poisson distribution has a similar position among the discrete distributions, if these are considered from an asymptotic standpoint. Further details see in some textbook.

4.2 Probabilistic Inference

Statements from data sets are possible more easily and better, if assumptions can be put basing their genesis plausibly or scientifically. If these assumptions lead to a *probabilistic model,* then the statements can be obtained as conclusions from the given data in the framework of this model. A probabilistic model can be formulated as a (random) experiment by fixing experimental conditions, which are taken for essential. As was mentioned already in the beginning of the chapter, simultaneously all other influencing facts on the result are explained as the effect of chance. In such way a random variable, mostly a one-dimensional one X, is defined. If its distribution is known *precisely*, then, on the given experimental conditions, *all available information* on the experimental result is specified, further statements are not possible. A near at hand example is yielded by the playing at dice: on the condition that the dice is fair, no more can be stated than that in the next throw each side has the same chance.

As a rule, the probability distribution of a practically defined random variable is *not precisely* known. From a model idea, e.g. the *type of distribution* may be deduced, for a measuring process perhaps the normal distribution and for statistical quality control the binomial distribution, but their parameters remain unknown. On the other hand some data material is given, which is related to the experiment and hence to the random variable and those data can contain information on these unknown parameters. With this *problem of parameter estimation* Subsect. 4.2.2 will be concerned.

Another problem considers the case that an experiment is given, the random variable of which obeys a distribution likewise given. By means of the given data material it should be decided, whether this is *actually* so. Such cases occur especially in statistical quality control and in process control. Problems of this kind are treated in Subsect. 4.2.3.

But first of all the important problem of application is to be considered: how a data material is to be gathered or which qualities it has to have so that it can answer the questions with respect to the distribution of the random variable. The central notion for this purpose is the *sample*.

4.2.1 Samples

Methods of *mathematical statistics*, as probabilistic inference is usually called and implemented in common software tools, base on several assumptions. If these assumptions are *not* fulfilled, then the statements obtained can be doubtful, even misleading and absurd. Therefore they should be presented rather detailed in this subsection.

A single special result of a trial or the value, which a random variable X has assumed *really* in a concrete case, is called a *realization*. When, e.g., a 4 was thrown with a dice or the result 17.3 was obtained in a measurement influenced by chance, so these are examples for realizations of throwing a dice or of the random variable modelling the measurement.

A number of such realizations $x_1, x_2, \ldots, x_n$ of a random variable X is called a *concrete sample*.

Because a statistical procedure should be usable *generally* in any possible case, *all possible* concrete samples must be considered. For this purpose they are explained as points $\mathbf{x} = (x_1, x_2, \ldots, x_n)$ in an n–dimensional space, the *sample space*. In order to indicate the probability that a certain sample $\mathbf{x_0}$ will occur or will lie in a certain domain of the sample space, the random vector $\mathbf{X} = (X_1, X_2, \ldots, X_n)$ is introduced. The concrete sample $\mathbf{x}$ (consisting of n realizations of X) is now interpreted as a single realization of this vector $\mathbf{X}$. Also, x_1 is a realization of X_1, x_2 of X_2, ..., x_n of X_n. Since all realizations x_i should come from the *same* random variable, all the X_i must have the *same* distribution, hence the X_i are called *copies* of X. Up to this point the proceeding is only a change in the line of sight to use the thought habit of the n-dimensional space in order to simplify the way of speaking.

If, however, the probability distribution of $\mathbf{X}$ should be computed, from which the desired probability values for the samples can be obtained, an *additional* assumption is necessary to deduce the distribution of $\mathbf{X}$ from that one of X. The mathematically *simplest* such assumption is that of the *independence* of the components X_i of the vector **X**. Then, namely, the distribution of **X** can be obtained by multiplication of the distributions of X_i. Let be, e.g., given the densities $f_i(x_i)$ of the corresponding random variables X_i for $i = 1, 2, \ldots, n$, then the density $f(\mathbf{x})$ of the distribution of **X** is

$$f(\mathbf{x}) = \prod_{i=1}^{n} f_i(x_i) \,. \tag{4.27}$$

A vector **X** of n *independent* random variables $X_i; i = 1, \ldots, n$, all with the *same* distribution, is called a (mathematical) sample. The number n is the sample size. The random variable X, which supplies the sample, is called the *population* (with respect to the sample).

Naturally, problems in statistics can be treated also without the assumption of independence, however, the procedures then necessary are essentially more complicated, as a rule, they demand other assumptions on the stochastic correlation of the components X_i or they are rather restricted in their meaningfulness. Considerations of random fields, e.g. in geostatistics, provide impressive examples (see, e.g. CRESSIE (1991)).

The important problem of *data quality* in application manifests itself here in the questions: What consequences have the specifications in the definition of a sample for the organization of data gathering? On what conditions can methods of mathematical statistics be applied meaningfully to given data? How must results be obtained to be regarded as a concrete sample, i.e. as a realization of a sample? These questions must be, naturally, answered *before* the consideration or the gathering of data is thought about!

These questions are discussed first for the case that the data (which should be considered as realizations of X) are *crisp* values. Data uncertainty (fuzziness) is additionally taken into account in Subsect. 4.2.4.

The sample is based by a certain random variable, i.e. by a random experiment with *fixed* experimental conditions; *all other* influences are explained as the effect of chance. The consequence of this determination are two necessary checking procedures:

Checking procedure 1: Are all the fixed experimental conditions *really* fulfilled in *every* trial (realization of the random variable)?

Hence the researcher is always recommended to get an idea of the situation in this respect, also if or even if he does not take the necessary measurements or observations himself.

If the question must be answered in the negative, then there is a risk that (some) results are recorded, which do *not* come from the considered random variable. In many a case these results lie outside the bulk of the other realizations and are called *outliers* in the software of data analysis. It is advisable *not* to depend on the "decision" of such software, which recommends simply to eliminate such results. It is, namely, absolutely normal that also erroneous values are hidden in the bulk and seperately lying values belong in fact to the considered variable. A scientifically based inspection of the results is recommended in any case, which should end in an individual decision on each single realization. If there is a possibility given to repeat situations with outlying values, then one should use it. If such values are confirmed in that way, then they can open new insight into the practical problem.

Moreover, it can happen that only realizations from a certain bounded domain can be recorded, whereas also realizations outside of this domain occur but remain undetected. Examples for this situation are lifetime investigations, in which not all objects under test failed out before the end of the test time, and observation values, which are prevented by control actions during the investigation (e.g. by switching on a supplementary device when approaching a dangerous domain). In such cases the samples are *truncated*. This can be detected, as a rule, by a scientifically trained eye in a consideration within the framework of data analysis. Then suitable measures are to be taken either to neutralize this truncation by assumptions on the basing distribution or to take this into account in the conclusions.

Finally, situations are to be considered, which can even depreciate the data for a probabilistic inference. If, namely, the employee, instructed to measure or to observe the realizations, is interested in *what* values he records (e.g. because his income or prestige depends on this) or he is apathetic against what he is recording (e.g. because care is neither demanded nor appreciated), then the *data might be manipulated*. The conclusions drawn from such data are no longer objective, sometimes not even meaningful. A sound scepticism in too good or senseless results likewise is always advised.

Checking procedure 2: More difficult to realize, but at least likewise serious the frequent case is found that *besides* the fixed experimental conditions *further* conditions are *fixed* in all trials of the obtained sample, i.e. for all the realizations. This conditions become automatically part of the experimental conditions, which define the population, and *all* conclusions drawn from the sample then are valid *only* on these additionally fixed conditions. This is, obviously, a *restriction* for the assumed effect of chance, which is not in the interest of the researcher in many cases. Known situations, in which such cases occur, are investigations with unintentional restrictions, e.g. selection of departments in an enterprise, choice of saison, day in the week and time of day in business management investigations; selection of the location for field tests in agricultural research; selection of manpower in engineering investigations, where experience and care is of importance. In order that the conclusions are valid for all possible objects (departments, seasons, days in the week, times of day, field locations, manpower) it must be taken care that the influence of them is "by chance", or, as it is frequently called in mathematical statistics, *at random*. This is effected in practical cases by including all possible objects into the investigation in a *random manner*, this is called *randomization*.

Examples for application of such procedures are known from many fields: experimental design for agricultural research, selection of objects for statistical quality control, sampling for the analysis of debris, process control, surveys among customers or voters. All possible objects are imagined to be numbered in some way and then the necessary number of random numbers are produced by a *random number generator* (with equal distributed digits). The objects with the realized numbers are selected for inspection or analysis. With this procedure it is guaranteed that each possible object has the *same* chance to be selected (see e.g. OGAWA (1974)).

Finally the definition of the sample demands that the realizations should be obtained *independently*. In a concrete case this means that any obtained realization x_i must not have any influence on what values the other realizations x_j with j different from i will have. The fulfilment of this demand can be checked, e.g. by consideration of the procedure for obtaining the realizations. Is there any possibility that in such a procedure running in time the results *already* obtained have any influence on those to be *still* obtained? For instance, there is a temptation for the employee instructed to take a gathered sample from debris during the gathering up of such a sample to "correct" the whole sample by intentionally selected further material from the debris. A further strategy for checking the demand of independence uses software for data analysis for finding correlation and functional relationships. One has, however, to apply a stricter standard to the results for their relevance, because it would be always possible to find *any* weak connections, but which will be originated "by chance" and can be quite compatible with independence.

The random selection of objects in more complex situations and with the claim that, on the one side, only relatively few objects are necessary for their intended analysis, and on the other side, simultaneously the whole variability of all possible objects is considered and reflected, is the subject of a very elaborated *sample theory* (see e.g. COCHRAN (1957)).

The sample vector $\mathbf{X}$ is the starting point as well as the basis of the whole probabilistic inference or mathematical statistics.

The *histogram* and the *empirical distribution function* (see textbooks for detail) are derived from the concrete sample and supply a first rough picture of the distribution (of the single probabilities or of the density) or of the distribution function itself. For increasing sample size n the approximation to the corresponding characteristics of the distribution becomes better and better (law of large numbers). For small n they are but little meaningful, one should, in this case, refrain from conclusions.

4.2.2 Parameter Estimation

One of the two basing problems of mathematical statistics is the estimation of unknown parameters of the distribution. Starting point is the assumption that the *distribution type* of the population is already known. The notation "distribution type" is understood mathematically by the notion *distribution family*. Such a family consists of all the distributions, the mathematical representations of which are distinguished only by different values of parameters.

For illustration here the well known family of normal distributions

$$f(x; \mu, \sigma^2) = \frac{1}{\sigma\sqrt{2\pi}} \exp\{-\frac{(x-\mu)^2}{2\sigma^2}\} \qquad (4.28)$$

with the parameter domain

$$\Theta = \{\vartheta = (\mu, \sigma) : -\infty < \mu < \infty; 0 < \sigma < \infty\} \qquad (4.29)$$

and the family of Poisson distributions

$$P_\lambda(X = k) = \frac{\lambda^k}{k!} \exp(-\lambda) \qquad (4.30)$$

with the parameter domain

$$\Theta = \{\vartheta = \lambda : 0 < \lambda < \infty\} . \qquad (4.31)$$

will serve as examples. This is the form, in which the distribution types are usually represented in textbooks and software.

The *main assumption of estimation theory* consists in that there is always a special parameter value ϑ^* to which the population of the given sample belongs. The fulfilment of this assumption is formulated in the wording: the distribution family is a *true model*. If this assumption is *not* valid in the concrete case, then the result of the estimation may be possibly meaningless.

The problem of estimation arises, because this *true* ϑ^* is *unknown*. An *estimation procedure* must assign a corresponding suitable estimation value to every possible sample from the population, this assigned value should act as an approximation of the assumed true one. Hence every element $\mathbf{x}$ from the sample space must correspond to an element $\hat{\vartheta}$ of the parameter domain, not necessarily one-to-one. This is called also, rather graphically, a *mapping* of the sample space into the parameter domain:

$$\hat{\vartheta} = \hat{\theta}(x_1, \ldots, x_n) . \qquad (4.32)$$

Before its realization the sample is a random vector $\mathbf{X} = (X_1, \ldots, X_n)$, and hence also the estimation value is "at random", since it depends on the random realization of the (mathematical) sample. Accordingly the random variable $\hat{\theta}(X_1, \ldots, X_n)$ is considered as the (by the sample random) *estimation* of the unknown true *parameter* value ϑ^* and also called an *estimator* of ϑ and its realization will be called an *estimate* of ϑ. Because the estimation procedure must be applicable for *any* true parameter value, the star is omitted after having emphasised the importance of the existence of such a value.

For an example the family of normal distributions is considered. As an estimator of the expected value μ the arithmetic mean $\bar{X}_n = \frac{1}{n}\sum_{i=1}^{n} X_i$ is suitable, it tends for increasing n towards the expected value (see Subsect. 4.1.4).

With respect to other estimation procedures for different families of distribution functions it is referred to textbooks or to software documents.

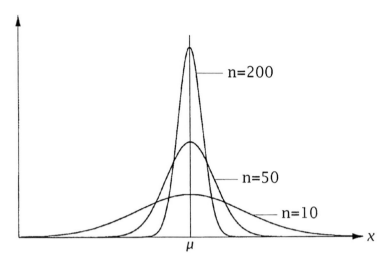

Fig. 4.3. The arithmetic mean as an estimator of the expected value for different sample sizes

Perhaps also *properties* of the recommended estimators are given, then their meanings will be of interest. Because the estimates in concrete cases are realizations of random variables $\hat{\theta}$ (see Fig. 4.3 for an example), these properties can be described only for those random variables.

If the *frequentistic* interpretation of probability is chosen, then *unbiasedness* of an estimator means that the estimates, which are obtained each by the procedure from a realization of the (mathematical) sample, in the long run, will scatter around the "true" value of the parameter (it is the expected value of the estimator). A *biased* estimator would show a *systematic* mean deviation from this true value. The *effectivity of an unbiased estimator* is measured by its variance, i.e. the expected square deviation from the true (and expected) value. The smaller this variance the higher is the effectivity of the estimator; the intensity of the scattering around the true value decreases.

The *consistency of an estimator* is mentioned in connection with its behaviour if the sample size tends towards infinity. If the estimator becomes more and more effective, or, in another sense, better and better (it contracts, so to speak, to the true value), then it is called a *consistent* estimator. Hence this property consists in an asymptotic statement.

Though all properties refer to special *sets* of randomly obtained estimates. With respect to more precise definitions and further properties see some textbook.

In a *single* concrete case *nothing* can be stated on the nearness of the obtained estimate to the true value.

This remaining uncertainty led to the concept of *confidence region estimation*. To every concrete sample $\mathbf{x}$ not only a single parameter value ϑ, but a whole region $B \subseteq \Theta$ is assigned. In the one-dimensional case $\Theta \subseteq (-\infty, \infty)$, which is considered in the following for the sake of simple presentation, this would be an interval I for practical reasons. This interval depends on the sample: $I(X_1, \ldots, X_n)$, and becomes so a *random interval*. Its realization for a concrete sample is then some fixed interval $I(x_1, \ldots, x_n)$. This interval is *assumed* to *cover over* the true but unknown parameter value ϑ. Naturally, this assumption can, in every single case, be either *true* or *false*. If this region estimation procedure is used permanently (frequentistic interpretation), then this assumption should be true as frequently as possible. The "fuzzy" formulation "as frequently as possible" must be made "precise" here within the framework of stochastics. Therefore the random interval is considered and the probability for the case that ϑ is the true value. This assumption is indicated by an index at P. Moreover the sign $\ni$ should emphasis that the set *before* is *variable* whereas the element *behind* is *fixed* (usually with $\in$ the *variable* element stands *before* and the *fixed* set *behind*). For the expected frequency of a correct decision (i.e. the true value is in fact covered up by the interval) a lower bound $1 - \alpha$ is given. Hence the (small) number α is the probability of a *wrong* decision, i.e. that the random interval does *not* cover over the true value. A random interval $I(X_1, \ldots, X_n)$ is called a *confidence region estimation* or here shortly *confidence interval* to the confidence probability $1 - \alpha$, if

$$P_\vartheta(I(X_1, \ldots, X_n) \ni \vartheta) \geq 1 - \alpha \tag{4.33}$$

is valid for all ϑ from the parameter domain Θ. The length of the interval is then a measure of the *accuracy* (generally the measure of the region)   and the probability $1 - \alpha$ gives obviously the *certainty* of the estimation. An example should explain the connection.

A normally distributed random variable X with the parameters μ and σ is considered. The parameter σ, the precision of the measuring procedure, let be known. The parameter μ, the true measurement, is to be estimated: $\mu = \vartheta$. Let denote $u_{1-\alpha/2}$ the $(1 - \alpha/2)$-quantil of the standardized normal distribution, i.e.

$$F_Z(u_{1-\alpha/2}) = \frac{1}{\sqrt{2\pi}} \int_{-\infty}^{u_{1-\alpha/2}} \exp\left\{-\frac{x^2}{2}\right\} dx = 1 - \alpha/2 \tag{4.34}$$

for the standardized normally distributed random variable Z (with $\mu = 0$ and $\sigma = 1$). (In general a *quantil* u_β indicates the point on the x-axis, where the probability $P(X \leq u_\beta)$ of a continuous random variable X reaches the value β exactly.) Then a good confidence interval to the confidence probability $1 - \alpha$ is recommended as

$$\bar{X} - u_{1-\alpha/2}\frac{\sigma}{\sqrt{n}} \leq \vartheta \leq \bar{X} + u_{1-\alpha/2}\frac{\sigma}{\sqrt{n}} \ . \tag{4.35}$$

The length of this interval as the accuracy of the estimation is

$$l(I) = 2u_{1-\alpha/2}\frac{\sigma}{\sqrt{n}} \tag{4.36}$$

and depends obviously on the precision σ of the measuring procedure used and the number n of measurements. Moreover the certainty $1 - \alpha$ influences the length via the value $u_{1-\alpha/2}$. From this situation some practically important conclusions can be derived. The accuracy of the estimation is proportional to the precision of a (single) measurement. The number of repeated measurements effects a shortening, hence it increases the accuracy, but only by its root. Therefore it is *not* worthwhile, to repeat the measuring procedure very often, especially if this costs time or money. From a certain number on there is no longer a remarkable increase in accuracy. Finally an increase in certainty of the estimation by $u_{1-\alpha/2}$ effects a lengthen of the interval and hence a decrease in accuracy. This means that there exists something like an *indetermination relation*. From a given sample one can obtain either a rather accurate estimation with moderate certainty or a rather rough estimation with high certainty. By the choice of α, the risk for an error with respect to the covering up of the true parameter, one decides on the compromise between certainty and accuracy. Obviously, however, in a single concrete case it remains further *uncertain* whether the given concrete interval covers up the true parameter, but one has by α at least an assessment of the faced risk when using this method in the long run.

In general the length of the confidence interval itself is *at random* and such a clear presentation as above is no longer possible. The main statements on the connection between accuracy and certainty remain but valid in this general case.

4.2.3 Testing of Hypotheses

The second basing problem of mathematical statistics considers the case that for a situation a probabilistic model is already found, which is assumed to be true. The model is represented by a random variable, the distribution of which is *accurately known*. By a sample it is to be decided, whether the assumption on the truth of the model is (still) valid or whether there are founded doubts against this assumption.

Two classical examples will be mentioned here.

A lot of low mass-produced items, for which a tolerated proportion p_0 of rejectives was agreed, is to be tested, whether this condition is fulfilled. This is a known situation in quality control.

The number X_n of rejectives in a sample of n items chosen at random obeys, as is well known, a *binomial distribution* with the expected value parameter np, with p as the rejective proportion of the lot. The supplier kept the agreement, if the proportion of rejectives is not larger than p_0: i.e. if $p \leq p_0$. This *hypothesis* is confronted with the result of the inspection, the number x of rejectives in the sample. If this result is only little probable if the hypothesis is true, then this hypothesis should be rejected. In practical cases this leads to consequences against the supplier, who is now suspected of breaking the agreement. Because the result of the sample is influenced by chance (the random selection of the items), also rather inprobable results are possible, even if the hypothesis is true. Hence an *always correct decision* is in principle *impossible*.

For the consignee of the lot there is a risk with each of his both options: To reject the hypothesis, although it is really true (only the result of the concrete sample was unfavourable). This leads to problems with the supplier. To take the hypothesis for acceptable after the sample, although it is false (there are essentially more rejectives in the lot than agreed). This leads to a loss, when the lot is used.

A solution of this dilemma is possible by economic considerations, in which the two kinds of loss are specified quantitatively and weighed up, e.g. by their expected values (see e.g. UHLMANN (1970)).

But usually only the *error of first kind* is considered, when the lot is complained, although it is according to the agreement. This is felt more embarrassing than the *error of second kind* , when the lot is accepted, although is has a higher proportion of rejectives than agreed. But this can be detected, when the lot is used and can be considered in further business connections.

Another classical case of testing a hypothesis is the plain process control. The characteristic value of a product is allowed to vary within a given tolerance interval. It is assumed that the corresponding variable obeys a normal distribution, where the probability outside the given tolerance interval corresponds to the tolerated proportion of rejectives. During the production time of many product items the parameter of the normal distribution can change, e.g. by running out of adjustment or by worsening of the process characteristics determining the product quality. The given parameter values of the product distribution above, perhaps with small surrounding intervals, form the *hypothesis* in this case. By regular samples from the running production process it should be decided, when the process should be stopped for a new adjustment or a new (partial) equipment. Here the decision dilemma manifests itself in the two possible wrong decisions:

The process is stopped, although the sample has, at random, included unfavourable products (false alarm). The loss consists here in an (avoidable) loss of production during the stoppage.

The process is *not* stopped, although it had been necessary, but the sample was still favourable. The loss consists here in an (avoidable) higher proportion of rejectives by continuing the production (overlooked running out of tolerance).

In both the cases just mentioned as also in other cases of application from the result of the sample a value of a *test variable*, as it is called, is computed, the accurate distribution of which is known on the assumption that the hypothsis is true. If the computed value is within a *critical domain*, as it is called, then the hypothesis is rejected. A rejection or a nonrejection of a hypothesis on the basis of a sample says *nothing* on its truth. The hypothesis can be true or false in any case.

If the rejection of a true hypothesis is considered as the more embarrassing wrong decision, then the *error probability* α is specified, which should bound from above the probability of such an event. This α determines then the extent of the critical domain for the test variable. Its position will also be influenced by considering alternative hypotheses, as they are called. More details see in some textbook.

In a single case, as with confidence estimates, no statement is possible, whether the decision on the hypothesis was right or wrong. Only when using the test procedure in the long run, statements become possible, from a frequentistic standpoint, on the *frequency* of wrong decisions. Hence the decision situations above are classical examples of such situations.

Application of the test conception to more or less unique events, as it is the case frequently in natural sciences, in economics or in medicine, does not mean a thing with the frequentistic interpretation. On the other side, the subjectivistic interpretation of probability leads to formulations, with which the applying sciences would have some problems: "The odds are 80 : 20 that this hypothesis is true." At present in such situations the probability is determined that the hypothesis is just rejected after the obtained sample, and it is called the *P-value* (see e.g. LEHMANN (1986)) This number should give an idea how strongly the data contradict the hypothesis. Because the final decision on the hypothesis (rejection or acceptance) is left to the user in considering the computed P-values, actually nothing can be said on the risk of such a decision. Nevertheless the problem is treated further, on certain conditions, for various questions of interpretation and some extensions of the concept see the literature cited in LEHMANN (1986).

4.2.4 Problems with Imprecise Data

In textbooks and software documents it is assumed, as a rule, that the considered realizations are given exactly, as real or natural numbers. The case that the realizations are given as other symbols, e.g. as characters, is classified into this framework in that then only frequencies are included into the statistical analysis.

By handing over to a computer the data are "precised" into *computer numbers*, in which the observation impreciseness is *no longer* reflected (e.g. by the number of given digits in the original data). These *pseudo-exact data* now undergo the numerically formulated statistical procedures, e.g. to calculate an estimate of a distribution parameter. Hence the results of the computation are additionally objects of the *procedure impreciseness*. Because the computers give their outputs, as a rule, with a fixed number of digits (correspondingly also in floating point calculations), the user is *unable* to assess the influence of the input data "precising" and the imprecision inherent in the procedure for making sense of the number of obtained digits. To avoid such calamity textbooks and software handbooks construct simple and clear numerical examples without any practical background, "all clear cases", even if they are sometimes practically "couched".

In practically application it is really important to know how numerically precise the results of a probabilistic inference turn out and how observation imprecision and data "precising" effect to the result.

If a direct pursuit of impreciseness and an analysis of procedure imprecision seem impossible or inappropriate, then it remains advisable to get at least an idea on this circumstances.

A very simple possibility to do so consists in the following: the given data are varied several times slightly within the framework of observation or measurement precision, e.g. by randomly varying the last given digits within their rounding intervals, and then each time the results of the computation are recorded. This can be very simply managed on the computer and should belong already to the software standard. By a visual inspection of the so found results one can see how many digits remain stable and the result to be used can be rounded accordingly. Especially with confidence estimations rounding should be effected always outwardly; e.g. an estimating interval must not get smaller by rounding, because in covering one might keep on the safe side and *not* lower the probability of covering.

In testing numerical reliability plays a role especially, if the value occurs in the neighbourhood of the critical value. In this case the temptation is insistent to change the error probability a posteriori to cause a "clear" situation. This is, however, questionable, if one has adopted the frequentistic interpretation of hypothesis testing. For the P-value approach this does no matter, because the decision on the hypothesis is effected individually in considering this computed value rather roughly.

If one is in the position to specify the fuzziness of each single element of the sample, e.g. by determining a *fuzzy number* for each single pseudo-exact datum, then one can assess the influence of at least this kind of uncertainty on the test value. The extension principle of ZADEH yields then a fuzzy number as the result of the inference (VIERTL (1996)).

Let be (see (4.32)) $\hat{\vartheta} = \hat{\theta}(x_1, x_2, \ldots, x_n)$ the estimate for the parameter ϑ from the concrete sample $x_1, x_2, \ldots, x_n$. Having specified for each component of the sample a corresponding fuzzy number $A_1, A_2, \ldots, A_n$ over the real axis, then one obtains by a near at hand generalization of the extension principle (3.33) the fuzzy set Θ with the membership function $\mu_\Theta(\vartheta)$

$$\mu_\Theta(\vartheta) = \sup_{\vartheta = \hat{\theta}(x_1, x_2, \ldots, x_n)} \min_i \mu_{A_i}(x_i) , \qquad (4.37)$$

on the assumption that the fuzziness of the single sample components do not influence each other, in the language of fuzzy set theory: that they are not interactive.

Obviously, the procedure impreciseness is left out of consideration, which can be much higher in solving the optimization problem above than in computing the initial estimate.

With respect to an analogous treatment of the problem of confidence region estimation, in which a *fuzzy* estimating region is computed, it must be noted that in fact the core of the region represents a confidence estimation of the given confidence probability, but that a meaningful transition of the concept of *covering* the unknown parameter value by the *fuzzy* region is still to be found. For instance the question is to be answered: What about the characteristics of the fuzzy confidence regions found by the extension principle among *all possible* fuzzy estimating regions and how are the properly fuzzy parts of the regions are to be evaluated in the sense of the frequentistic interpretation of covering probability?

Nevertheless, this approach led already to practicable procedures even for rather sophisticated statistical problems (see e.g. FILZMOSER/VIERTL (2004)).

An idea of the influence of the observation fuzziness on the statements of mathematical statistics is yielded by the extension principle at any rate. A direct treatment of the fuzziness in the sample, which is useful in "very fuzzy" data, was presented by KRUSE/MEYER (1987) (see also Subsect. 5.3.1). Starting from their concept of fuzzy confidence regions GRZEGORZEWSKI (2000) constructed tests with fuzzy data.

4.3 Bayesian Theory

Probabilistic inference, as was presented in the preceding section, usually starts at the point that for information on the parameter to be estimated *only* the sample obtained is available. Previous knowledge on the situation consists only in a reasonable assumption on the distribution type of the population.

There are situations, however, in which already "certain knowledge" on the value of the unknown parameter is also available. But this knowledge does *not* (yet) suffice the demands on an estimation with respect to accuracy and certainty. These rough ideas contain but information on the unknown parameter, the inclusion of which into the inference arouses some hope for its improvement.

The three essential sources for such *previous knowledge* are theories from the applying sciences, somehow more vague experiences, and finally data material from investigations in the past.

Scientific theories indicate mostly that certain subdomains for the parameter are impossible for factual reasons, although they are not excluded by the statistical model. Frequently these bounds can be adjusted in suitable manner into the statistical procedure.

Available experience cannot be expressed, usually, in such a sharp form. Therefore they are frequently formulated as probabilistic statements. They are specified as *a-priori distributions* for the unknown parameter. Because these are, as a rule, *subjective* determinations, the comments from Subsect. 4.1.2 on subjective probability are valid also here: They are rough findings, which do not claim high numerical accuracy and factual reliability. Hence recently fuzzy specifications of a-priori distributions are introduced and used to generalize BAYES' theorem (see VIERTL/HARETER (2004)).

Finally in certain cases data material from investigations in the past can be used. This can be done either by estimating an a-priori distribution of the parameter from the conditional distribution of the (mathematical) sample combined with the empirical distribution of the concrete sample (see MARITZ (1970) for details). Or, heuristically, by surrounding the parameter value, estimated from the sample, by a distribution, which reflects the estimation accuracy and possible deviations from the conditions in the past. Sometimes the old data can be introduced as additional components into the present sample, but a sound scepticism is advised, first of all, if the conditions of their occurrence thence are only scarcely known.

The possibilities to formulate a-priori distributions form the starting point of the BAYESIAN theory or statistics, as it is called, which will be the subject of the next subsection. A rather reasonable presentation also of the problems in application of this approach one can find in BERGER (1985).

Application of BAYESIAN theory in practice is always connected with the problem of interpretation. It fits into the framework of frequentistic philosophy, if the parameter itself can be considered a random variable. Then the specified a-priori distribution can be explained as a (perhaps only heuristic) estimation of some "true" but unknown distribution with all the consequences of uncertainty from its origin, which are bequeathed to all statements derived from it. From another standpoint the a-priori distribution may reflect also the present state of knowledge with respect to a fixed but unknown value of the parameter. Then it is part of the subjectivistic philosophy and all statements derived from it demand for an appropriate interpretation. These problems one has to have in mind, when BAYESIAN theory is applied.

4.3.1 Bayesian Inference

Combining information from the a-priori distribution with that contained in the sample on the parameter is effected, in principle, by the BAYES' theorem (4.5). However, in the continuous case of an a-priori density $\pi(\vartheta)$ and a continuous population X with the distribution density $f_X(x|\vartheta)$, the dependence of which on the parameter ϑ is given explicitly, this formula has a more complicated form. The a-posteriori density $\pi(\vartheta|\mathbf{x})$ of the unknown parameter ϑ after the sample $\mathbf{x} = (x_1, \ldots, x_n)$ looks like

$$\pi(\vartheta|\mathbf{x}) = \frac{f(\mathbf{x}|\vartheta)\pi(\vartheta)}{\int_\Theta f(\mathbf{x}|\lambda)\pi(\lambda)d\lambda} , \qquad (4.38)$$

where $f(\mathbf{x}|\vartheta) = \prod_{i=1}^{n} f(x_i|\vartheta)$ is the density of the (mathematical) sample.

With the a-posteriori distribution a *predicative distribution* can be computed, which in the continuous case is given by its *predicative density*

$$f(x|\pi, \mathbf{x}) = \int_{\Theta} f(x, \vartheta)\pi(\vartheta|\mathbf{x})\mathrm{d}\vartheta \tag{4.39}$$

which unites the information on the statistical model, the a-priori density and the sample results and could be used to predict the behaviour of the random variable X in future.

From the a-posteriori distribution of the parameter estimating values and estimating regions can be obtained. E.g., the expected value of this distribution is used and called *a-posteriori* BAYES *estimate*.

The confidence region $I(X_1, \ldots, X_n)$, which should cover up the true parameter with probability $1 - \alpha$, is replaced by a HPD region (**highest a-posteriori** – **density** – **region**) $\Theta^* = \{\vartheta : \pi(\vartheta|\mathbf{x}) \geq k(\alpha)\}$. This region is defined as the set of all parameter values ϑ, for which the a-posteriori density $\pi(\vartheta|\mathbf{x})$ reaches at least the value $k(\alpha) > 0$. This $k(\alpha)$ is a constant, *as large as possible*, for which simultaneously

$$\int_{\Theta^*} \pi(\vartheta|\mathbf{x})\mathrm{d}\vartheta = 1 - \alpha . \tag{4.40}$$

The elements of Θ^* must have a certain minimum probability (here given by the density value) and the whole region must have the demanded confidence probability. Opposite to the confidence region (see (4.33)) which is assumed to *cover up* the unknown true parameter value ϑ, the HPD region is interpreted as a region, which *contains* the random variable Θ with probability $1 - \alpha$.

Sometimes the problem is considered within the framework of *decision theory*, in which the true unknown parameter is regarded as a state of a *nature*. As the *nature* all that is understood what the *decider*, the "antagonist" of the nature, cannot influence. The estimation problem is looked at to be an experiment of the decider to meet the true parameter value from the result of the sample. Let be ϑ the true value and $\hat{\vartheta}(\mathbf{x})$ the value estimated from the sample; then the decider is penalized by a loss of $L(\vartheta, \hat{\vartheta})$. Because he uses the estimating procedure in the long run his estimates are realizations of the estimator, a random variable $\hat{\theta}(\mathbf{X})$, a function of the mathematical sample $\mathbf{X}$. In this way the loss becomes also a random variable

$$L(\vartheta, \hat{\theta}(\mathbf{X})) . \tag{4.41}$$

In the permanent use of the procedure the *average* loss is of interest, i.e. its expected value, which is called the *risk R*

$$R(\vartheta, \hat{\theta}) = E_{\mathbf{X}} L(\vartheta, \hat{\theta}(\mathbf{X})) . \tag{4.42}$$

The dependence of the risk on the unknown true parameter ϑ can be removed, if one has an a-priori distribution $\pi(\vartheta)$. The expected value of R with respect to this distribution is called BAYES' risk

$$\rho(\pi, \hat{\theta}) = E_{\pi} R(\vartheta, \hat{\theta}) . \tag{4.43}$$

The best estimator in this sense, with respect to the loss function L and the a-priori density π, is that, which *minimizes* the BAYES' risk.

For application the most important mathematical result of this approach is the statement that an estimate according to this best estimator can be obtained, if for the concrete sample $\mathbf{x}$ the a-posteriori expected loss is minimized:

$$E_{\pi(\vartheta,\mathbf{x})}L(\theta,\hat{\vartheta}) = \min_{\hat{\vartheta}} , \qquad (4.44)$$

where θ is here the random variable, with respect to which the expectation is to be computed. For a square loss function $(\vartheta - \hat{\vartheta})^2$ this is even the expected value of the a-posteriori distribution.

The decision theory approach, here demonstrated by the example of point estimation, can be applied to more general decision situations.

Starting from the standpoint of application of mathematics decision theory offers the advantage that many statistical procedures can be provided with practical evaluations, whether by financial quantification of the risk or by assessment of the value of previous knowledge (see e.g. FERGUSON (1967)).

These advantages, however, cannot be used in many cases, because besides the specification of the a-priori distribution π first of all it is the determination of a realistic loss function what makes the problem in practical cases. In using of aspects from decision theory within mathematical statistics one goes back, as a rule, to mathematically well manageable types of functions as e.g. square loss functions as mentioned above. For multidimensional parameters, as in regression (see Sect. 7.1), matrix-valued loss functions are rather popular. In all these cases the loss function gives only a qualitative impression of the consequences of possible wrong estimations and hence theoretical insight into the behaviour of the estimations.

4.3.2 Hierarchical Inference and Robustness

The problematic *exact* determination of the a-priori distribution, how it is necessary for a use of the BAYES' theorem and the other procedures mentioned in the preceding subsection, can be avoided: For the a-priori distribution only a family of distributions is chosen, the parameter of which is replaced by another family of distributions and so forth. This proceeding is called a hierarchical BAYESIAN approach. Certainly, this hierarchy is stopped on a suitable level, as a rule on the second one, because the influence of the determinations, if they are taken meaningfully from the situation and problem environment, will decrease with increasing level. Specifications on a higher level can be easily led back to those on the lowest level. Let be, e.g., $\pi_1(\vartheta|\lambda)$ an a-priori density with the still free parameter λ and $\pi_2(\lambda)$ a density of second level for this λ; then the resulting a-priori density for ϑ is obtained by

$$\pi(\vartheta) = \int_\Lambda \pi_1(\vartheta|\lambda)\pi_2(\lambda)\mathrm{d}\lambda . \qquad (4.45)$$

This approach can become a tentation, it leads easily to imaginations playing away from the given problem. For BAYESIANS, however, this is an important argument for a *universal* usefulness of their methods.

In recent times the hierarchical approach seems to have lost a lot of its importance, since now computation methods are available to handle *arbitrary* a-priori distributions. This approach, however, is not only a way to *determine* an a-priori distribution but also to *reflect* the present state of previous knowledge.

If on the parameter *too little is known* to determine a plausible a-priori distribution, then as a rule it is recommended to choose a *non-informative* distribution, which prefers non of the possible parameter values. Very frequently this will be a uniform distribution or, if the domain for the parameter is infinite, an improper uniform distribution, being constant over the domain and hence no longer a proper distribution.

In the case of multidimensional parameters the immediate determination of an a-priori distribution is frequently very or even too complicated. If one can assume that the parameter components, considered as random variables, are independent of each other, then the a-priori distributions can be determined componentwise and afterwards be multiplicated for the vector distribution. If this is not possible or not for all components, then frequently conditional distributions can be determined, which can be combined according to the usual rules.

On another side the statements obtained by the BAYESIAN approach depend very frequently only little on the type of the a-priori distribution. It suffices mostly, especially on higher levels, to fix the parameters approximately. This *robustness* is very pleasant for application. Robustness is defined, as generally in mathematics, as that the conclusions are not changed essentially, if the assumptions and conditions are varied within a reasonable framework, whether with respect to the a-priori distribution or to the loss function or, finally, to the distribution type of the sample. Corresponding investigations of sensitivity can be carried out, e.g., by repeating the procedures under slightly changed conditions with respect to distribution type and parameter values. On the other hand, by a suitable (arbitrary) choice of the a-priori distribution and of the loss function the given sample can lead to any desired conclusion, it may be even absurd.

Especially for linear models (see Subsect. 7.1.5) the robustness of the BAYESIAN approach will be used advantageously with respect to the distribution types.

A further advantage of BAYESIAN methods is their usefulness in *sequential procedures*. Starting with an a-priori distribution a sample of small size is realized from the population and the a-posteriori distribution is computed. If the so obtained statements, e.g. for parameter estimation or for prediction, do not suffice with respect to accuracy and certainty, then the a-posteriori distribution is declared to be the *new* a-priori distribution and a further sample of small size is realized and the procedure is repeated. The sample size depends on the given situation, e.g. on the costs and possibilities of a short-term re-alization, and can even mean *only one* realization of the population in each step. Therefore the sequential BAYESIAN procedures are so popular in *learning theory* and a standard in probabilistic neural networks (see Subsect. 6.1.5).

For a simple implementation of such sequential procedures it is sometimes recommended to choose a type for the a-priori distribution, which is *conjugate* to the distribution of the sample, the parameter of which are suitably adapted. Let be $F(\mathbf{x}|\vartheta)$ the distribution of the sample, then the family of *a-priori distributions* $\pi(\vartheta; \lambda)$ is a conjugate family to $F(\mathbf{x}|\vartheta)$, if the a-posteriori distribution $\pi(\vartheta|\lambda)$ belongs again to this family. If $\pi(\vartheta|\lambda_0)$ was chosen as the a-priori distribution, then there exists a λ_1 for the a-posteriori distribution of the same type. In this way in the next step of the sequential procedure one has the same starting situation, only with another value for λ. However, sequential procedures, e.g. in BAYESIAN learning theory, are *not* restricted to the concept of conjugate distributions.

4.3.3 Numerical Problems

As already mentioned in Sect. 4.1 a-priori distributions are not very precise, i.e. a high precision in computing a-posteriori distributions and of conclusions derived from them does not make sense. This is especially important, if the determination of an a-posteriori distribution can be obtained only numeri-cally. The integrals occurring in BAYES' theorem, particularly if they have multidimensional domains, can be computed frequently only approximately by Monte-Carlo simulation. Here a few essential digits may suffice, what can reduce the efforts essentially and keep the *real* impreciseness transparent.

Sometimes the assumption is introduced that the distribution of the sam-ple and of the a-priori distribution are *independent* of each other, because this simplifies their joint handling. Strictly speaking, this is nearly never fulfilled, because by the choice of the model and the consideration of the data for this purpose it is just defined what should be meant by the parameter ϑ. Moreover, by a preceding data analysis imaginations are inspired on the possible values of ϑ. Such assumptions on independence are valid at most approximately and hence to be considered in this manner.

A further problem in using the BAYESIAN approach is the subjectivity in specifying the a-priori distribution (and perhaps the loss function). One sees easily, however, that also the choice of the model is a *subjective* decision, even if it is sometimes, seemingly objectively, left to an automatic procedure from the software. But already the possibilities implemented are a result of a subjective decision (by the software producer). Finally it should be pointed out that already in defining "randomness" (see Subsect. 4.1.1) it was stated that by the choice of the essential experimental conditions it was decided (subjectively) what should be considered as the effect of chance in the given case.

Closing this subsection the question will be examined what should be done if between the imagination on the position of ϑ, given by the a-priori distribution, and the results $\mathbf{x}$ of the sample obtained, there are discrepances or perhaps even contradictions. A first hint is always to confront conclusions derived from the a-posteriori distribution, e.g. the presented estimate, with the corresponding conclusions, which come *only* from the sample. Naturally, differences between corresponding conclusions are based by the a-priori information used additionally. If the discrepances are rather large, then one should trust more in the result of a sample of not too small size than in the a-priori distribution. At least, obvious contradictions are an important reason to check the specification of the a-priori distribution as well as the data quality (see Subsect. 4.2.1) of the sample.

5

Specification of Vagueness
of Statements on Sets

In contrast to the preceding chapters, where the existence of extensive software tools could be assumed, now some approaches for mathematical treatment of certain kinds of uncertainty are presented, for which this is not the case, at least not in the same extent. Hence the presentation is restricted here to the explanation of ideas and trains of thoughts in order to inform, what could be managed by such a software for practical problems, when the tools will be developed in the course of time.

5.1 Fuzzy Measures

5.1.1 The Idea of a Fuzzy Measure

A function assigning a value to every single *set* of a certain kind is called a *set function* or a *measure*. This is clear, if one considers the length of a distance or the weight of a bag of flour: it is *measured*. However, not only material measurements will make sense.

So, purely theoretically, a certain element x_0 of the universe U could be also described uniquely by specifying for *every single possible subset* of the universe, whether the element x_0 is contained in it or not. The set of all subsets of U is called its *power set* $I\!\!P(U)$, and the just specified mathematical object would be the *localization measure* g_{x_0}, which assigns to every single set $A \in I\!\!P(U)$ the value 1, if $x_0 \in A$, and 0, if this is not the case. This proceeding takes for granted that one does *exactly know*, where x_0 is located, and hence it is practically senseless. The approach becomes meaningful again in the moment, when the element x_0 of interest can be located *not yet exactly*, e.g. the cause of a disease, the culprit of a crime, or the affilation of a discovered fossil to a species. This gives cause for a *fuzzy description* of this element, not by a fuzzy set in U, but by specifying a corresponding degree of assignment for every one of the crisp sets of $I\!\!P(U)$. In this way one obtains the *fuzzy measure g*. For the function

$$g : I\!P(U) \rightarrow [0, 1] \tag{5.1}$$

it makes sense, obviously, to demand that

$$g(\emptyset) = 0 \quad \text{und} \quad g(U) = 1 \tag{5.2}$$

and moreover

$$A \subseteq B \Rightarrow g(A) \leq g(B) , \tag{5.3}$$

i.e. that the degree of assignment can not decrease, if the set is enlarged. Or, with other words, if $A \subseteq B$, then the statement "$x \in A$" is less certain than the statement "$x \in B$".

For finite universes these properties suffice already to develop a useful theory and application of such fuzzy measures. For infinite universes continuity with respect to set inclusion is required, i.e. if a series of sets growing in extent and lying each within the following one tends towards a limiting set, then the series of the corresponding values of the fuzzy measure converges to the value of the measure of this limiting set.

In Subsect. 4.1.1 is was mentioned that random events can be represented by corresponding sets. So, a *probability* measures the chance that a random variable realizes an event that corresponds to the set A, or, with other words, that the realization lies within A. Hence probability distributions yield special measures. Typical for such a *probability measure* is its *additivity* as expressed in formula (4.2):

$$A \cap B = \emptyset \Rightarrow P(A \cup B) = P(A) + P(B) . \tag{5.4}$$

This is, by the way, a property, which probability shares with all measures of mass and length. Because monotonicity follows directly from additivity, probability measures are special fuzzy measures.

With respect to probability, it can be interpreted also as a statement on the position of a *not yet localized element* of a set of possible elements (the universe U) for every possible set of the power set $I\!P(U)$. (Probability is usually introduced only for a subset of $I\!P(U)$, a sigma-algebra $I\!B$, however, this does not play any role for the present argumentation.)

Statements on the possible position of a not yet localized element in all possible sets on the condition of *uncertainty* form the subject of this chapter. Variability in the sense of the frequentistic interpretation of probability is, as is known, only *one* kind of uncertainty. In contrast with fuzzy sets, by which statements are expressed on a given case (description of uncertainty), by fuzzy measures statements are expressed on cases not yet occurred using fuzzy knowledge on the given situation. Hence probability measures can be considered as particular cases of fuzzy measures.

5.1.2 Forms and Properties of Fuzzy Measures

Already in Sect. 4.1 arguments were considered, which question the only use of probability for a description of uncertainty. Especially for a description of the individual state of knowledge on a given situation additivity is sometimes rather restrictive. Two events can, with respect to their *possibility*, seem to be quite equal, whereas their real *probabities* can differ to some extent. Another situation is met in the diagnosis e.g. of diseases: nobody can be sure that the list of possible diseases is complete and that the items on the list are all disjoint.

From monoticity as demanded by (5.3) it follows immediately that

$$g(A \cup B) \geq \max\{g(A), g(B)\} \tag{5.5}$$

and

$$g(A \cap B) \leq \min\{g(A), g(B)\} . \tag{5.6}$$

The limiting case in (5.5) was used by ZADEH (1978) for defining the *possibility measure* Poss:

$$\text{Poss } (A \cup B) = \max\{\text{Poss } (A), \text{Poss } (B)\} . \tag{5.7}$$

It should denote the *degree of possibility* that an element of interest not yet located is situated within the set forming the argument.

If the universe U is *finite*, then every possibility measure Poss can be defined by its values assumed on the single elements $x \in U$:

$$\text{Poss } (A) = \max_{x \in A} \pi(x) , \tag{5.8}$$

where

$$\pi(x) = \text{Poss } (\{x\}) \tag{5.9}$$

is the *possibility degree* of the element x. Then the function $\pi(x)$ is called a *possibility distribution*. Because of the reasonable condition Poss $(U) = 1$ the function $\pi(x)$ is normalized, i.e. there is at least one element with possibility degree 1. Hence the possibility distribution $\pi(x)$ has the property of a normalized membership function and can be explained as the membership function $\mu_B(x)$ of a fuzzy set B, more precisely, as the degree of possibility that a variable v assumes the value $x \in U$. The fuzzy set B is often called the *inducing* set and carried with when specifying possible degrees of some variable:

$$\text{Poss } (\{x\}) = \text{Poss } (v = x | B) ; \quad \text{Poss } (A) = \text{Poss } (v = A | B) . \tag{5.10}$$

So, one could ask for the degree of possibility that a good student (element of the fuzzy set B) will achieve an only sufficient result in a certain exam, i.e.

that he belongs to the crisp set A of those students, for which that exam is marked "sufficient".

If U is *infinite*, then such a possibility distribution need not exist. However, this existence is guaranteed, if the condition (5.7) is extended to infinite unions of sets. In practical application the existence can be always assumed and in (5.8) max can be replaced by sup. However, in case of infinite universes U possibility measures need not satisfy the condition of continuity, as mentioned in the preceding subsection, and hence then they are no longer fuzzy measures in the sense given above (see PURI/RALESCU (1982) for details).

The other limiting case, in (5.6), leads to necessity measures Nec, as they are called, which hence satisfy the condition

$$\text{Nec } (A \cap B) = \min\{\text{Nec } (A), \text{Nec } (B)\} . \tag{5.11}$$

The necessity measure indicates the degree that a non-located element of U is situated *necessarily* within the set forming the argument. This interpretation becomes clear when one considers that (5.11) is eqivalent to

$$\text{Nec } (A) = 1 - \text{Poss } (A^c) . \tag{5.12}$$

this reflects, at least for the function values 0 and 1 of unique membership, the intuitively clear idea that an event A must be *necessary* if its contrary A^c is *impossible*.

Naturally, via (5.12) and (5.9), also a necessity measure can be constructed by a given possibility distribution.

Since all considered argument sets here are *crisp* sets, it holds $A \cup A^c = U$ und $A \cap A^c = \emptyset$. For such sets one obtains some interesting properties of these fuzzy measures, which motivate their naming:

$$\text{Nec } (A) \leq \text{Poss } (A) , \tag{5.13}$$

because what is necessary must be even possible. Moreover, one has

$$\text{Nec } (A) > 0 \quad \Rightarrow \quad \text{Poss } (A) = 1 ; \tag{5.14}$$

which means: what is necessary to any degree, as small as it may be, that is absolutely possible, without any restriction; and, in the other direction

$$\text{Poss } (A) < 1 \quad \Rightarrow \quad \text{Nec } (A) = 0 , \tag{5.15}$$

what is not *absolutely* possible that cannot be necessary at all.

Obviously, these two measures are rather extreme in their properties. How they can be used nevertheless, especially in their interplay with fuzzy sets, can be found in some detail in DUBOIS/PRADE (1988).

Frequently the problem is merely to *weaken* somewhat the likewise hard condition for additivity of probability.

A very interesting suggestion is due to SUGENO (1974). For disjoint sets $(A \cap B = \emptyset)$ he suggests to replace formula (5.4) by the following connection formula

$$Q_\lambda(A \cup B) = \min\{Q_\lambda(A) + Q_\lambda(B) + \lambda Q_\lambda(A)Q_\lambda(B), 1\} \qquad (5.16)$$

which turns into (5.4) for $\lambda = 0$ in (5.4). In this way, by Q_0 probability is contained in the set of these λ-measures, as they are called. The conditions, put on a fuzzy measure, are fulfilled by these measures for all $\lambda > -1$. Further corresponding formulae for application can be obtained from the definition of fuzzy measures and the connection formula above. So one has for the complementary set

$$Q_\lambda(A^c) = \frac{1 - Q_\lambda(A)}{1 + \lambda Q_\lambda(A)} \qquad (5.17)$$

and for the union of arbitrary sets:

$$Q_\lambda(A \cup B) \qquad (5.18)$$

$$= \min\left\{ \frac{\left(Q_\lambda(A) + Q_\lambda(B) - Q_\lambda(A \cap B) + \lambda Q_\lambda(A)Q_\lambda(B)\right)}{\left(1 + \lambda Q_\lambda(A \cap B)\right)}, 1 \right\}.$$

The application one can imagine in the following manner: Certain important pairs of sets from the situation are considered and the measure values of the pairwise unions of these sets are specified using the measure values of the single sets according to the ideas of the applying scientist. Then a λ is computed, which approximately corresponds to these specified measure values. In this way, one obtains an impression, how strong the ideas of connection deviate, in the given case, from those for probability.

For the special case that the universe U is the real axis X, Q_λ can be defined by a function h, which shows the properties of a continuous distribution function of a random variable, i.e. being continuous monotonuosly non-decreasing with limits 0 and 1 (see SUGENO (1977)). Then for all intervals $[a, b]$ one has

$$Q_\lambda\left([a, b]\right) = \frac{h(b) - h(a)}{1 + \lambda h(a)}. \qquad (5.19)$$

5.2 Simple Inference with Fuzzy Measures

The most problematic but also most important task for the use of fuzzy measures with respect to inference is the *specification* of such measures, i.e. the determination of their values for different interesting sets. In the first subsection an approach to fuzzy measures is presented, which uses the vocabulary of probability theory to model the state of knowledge of some individual on a situation.

5.2.1 Specification of Partial Ignorance

As is known, a probability measure is given, if the probabilities are known for all possible events of a given field of events $I\!A$. Naturally, these probabilities can be computed from those ones for all elementary events. Often, in practical cases, however, probabilities can be specified meaningfully only for certain events of the field $I\!A$, because there is no more known yet on the basing practical problem. In some cases from this knowledge the yet lacking probability values can be computed according to the rules of probability theory, in other cases the given information does *not* suffice for this purpose. Possibly the first reference to such a problem can be found already in BOOLE (1854), who considered the case that for two events A, B only the probability values

$$P(A) = p \quad \text{and} \quad P(A \cap B) = q \tag{5.20}$$

are given. From these values for $P(B)$ only bounds

$$q \leq P(B) \leq q + 1 - p \tag{5.21}$$

can be concluded and hence *no unique* determination of this probability is possible.

This deficiency reflects the insufficient state of knowledge in this case: the *partial ignorance*, as it is called.

For *finite* universes U SHAFER (1976) presented an interesting concept, by which *fuzzy measures* can be constructed from such an insufficient state of knowledge. The total weight 1 is spread out over the power set $I\!P(U)$ of the universe

$$p | I\!P(U) \to [0, 1] \quad \text{with} \quad \sum_{B \in I\!P(U)} p(B) = 1 \tag{5.22}$$

(and not, as with probability, over the elements of U). Naturally in this distribution the empty set should obtain the weight 0. The set function p is called *basic probability assignment*. The sets with positive weights ($p(A) > 0$) are the *focal sets* of p. The set of all focal sets, so to speak the support of p, will be denoted by supp p. The pair (supp p, p) was called a representation of a *body of evidence* by SHAFER (1976). The weight $p(A)$ can be interpreted in different manner. It can be understood as a *remainder* of the probability $P(A)$, which – at the given state of knowledge – cannot be distributed further to subevents of A. On the other hand the value $p(A)$ is frequently considered as the *relative level of confidence* in A as a representation of the available information. It represents the "probability" that this information is described correctly and completely by $x \in A$.

The focal sets need not be neither disjoint, nor form a covering of U. Even U can be a focal set. Then $p(U)$ means that portion of confidence, which is owed to ignorance. Hence *total ignorance* is expressed by $p(U) = 1$. (Here it becomes clear, why one should question the advice to assume an equal distribution in the case of total ignorance with respect to a probability distribution.)

This interpretation of p rests on the assumption that the focal set A describes all possible positions, which the value of a certain variable can assume, e.g. A can be a fuzzy observation or measurement, perhaps recorded only verbally. In this context the information is called *disjunctive*, in such a sense that the actual value of the variable is unique. Hence the focal sets represent mutually excluding possible values of the variable.

The focal sets and their *evidence weights p* are specified, as a rule, *subjectively*; the approach was created even for this purpose. So it may easily happen that different experts specify differing bodies of evidence in the same situation. By DEMPSTER (1967) (see also DUBOIS/PRADE (1988)) a rule was suggested, by which *discrepancies* between different assignments can be reconciled to a certain degree.

For presenting the rule an intermediate step seems useful. Let denote by p_1 and p_2 the respective probability assignments of two experts. For all $A \in I\!\!P(U)$ the *formal product*

$$(p_1 \cdot p_2)(A) = \sum_{B \cap C = A} p_1(B) \cdot p_2(C) \tag{5.23}$$

is considered. The basic probability assignments for this product assume again values in $[0,1]$, but the sum over all $A \in I\!\!P(U)$ can be less than 1. The probability assignment to the empty set can be positive:

$$(p_1 \cdot p_2)(\emptyset) > 0 . \tag{5.24}$$

This reflects the *conflict* between the two assignments p_1 and p_2. If the case of *total conflict* $(p_1 \cdot p_2)(\emptyset) = 1$ is excluded, which seems to be reasonable, then by renormalizing $(p_1 \cdot p_2)$ a conflict reconciling basic probability assignment for all $A \neq \emptyset$ can be generated, which will be denoted by $(p_1 \cap p_2)$:

$$(p_1 \cap p_2)(A) = \frac{(p_1 \cdot p_2)(A)}{1 - (p_1 \cdot p_2)(\emptyset)} . \tag{5.25}$$

This is the announced DEMPSTER rule. The application of this rule will be somewhat *problematic*, if the conflict is very hard, the assignment by the product to the empty set reaches a remarkable value. Then the rule may be even instable, inasmuch as slight changes in the assignments of the single experts cause considerable changes of the assignment according to this rule. In such a case one should interpret the result as a signal to reconsider the situation and the experts again carefully.

In general, the probabilities of the events, $(P(A))$, themselves remain unspecified by a basic probability assignment p. It is only known that the probability $P(A)$ lies within an interval $[P_*(A), P^*(A)]$, where

$$P_*(A) = \sum_{B \subseteq A} p(B) , \tag{5.26}$$

$$P^*(A) = \sum_{A \cap B \neq \emptyset} p(B) .$$ (5.27)

So, the value $P_*(A)$ is computed by considering all focal sets B, which make the event A a *necessary* one (i.e. as a consequence), whereas for $P^*(A)$ all focal sets are taken into account, which make the event a *possible* one.

Moreover, there exists a duality relation between the values

$$P^*(A) = 1 - P_*(A^c) .$$ (5.28)

However, P^* and P_*, in general, are no longer possibility and necessity measures, respectively. This is only then valid, if the focal sets are nested, this case is then called the *consonant* case. More precisely, if for the focal sets it holds

$$A_1 \subset A_2 \subset \cdots \subset A_s ,$$ (5.29)

then the corresponding possibility distribution is defined by

$$\pi(x) = P^*(\{x\}) = \begin{cases} \sum_{j=i}^{s} p(A_j) , & \text{if} \quad x \in A_i; \quad x \notin A_{i-1} , \\ 0 , & \text{if} \quad x \in U \backslash A_s . \end{cases}$$ (5.30)

If, on the other hand, all focal sets are elementary events (respectively atoms) and hence *disjoint*, that is the *dissonant* case, then it holds obviously for all A of the power set

$$P_*(A) = P(A) = P^*(A) .$$ (5.31)

When the state of knowledge is expressed by a body of evidence it becomes clear that probability measures address precise but differentiated items of information, whereas possibility measures reflect imprecise but coherent items. So, possibility measures are useful for subjective uncertainty: one expects from an informant no very precise data, however, one expects the greatest possible coherence among his statements. On the other hand, precise, but variable data are usually the result of carefully observing physical phenomena.

As a rule, the state of knowledge is neither precise nor totally coherent, i.e. P^* and P_* are neither probabilities nor possibility or necessity degrees, respectively. Hence SHAFER (1976) called the measure P_*, defined by (5.26) for finite universes U, in the general case, the of A *degree of credibility* (or of belief)

$$\text{Cr} (A) = \sum_{B \subseteq A} p(B) .$$ (5.32)

It represents the *weight of evidence*, the *degree of confidence*, concentrated on A, i.e. on events, which have the occurrence of A as a consequence. Deducing from this measure, SHAFER defined by

$$\text{Pl}\,(A) = 1 - \text{Cr}\,(A^c) = \sum_{A \cap B \neq \emptyset} p(B) \tag{5.33}$$

the *degree of plausibility*, the degree of "understanding", which obviously conincides with P^* according to (5.27). It represents the weight of evidence concentrated on events, which make the occurrence of A a possible event.

Interpreting $\text{Cr}\,(A^c)$ as the degree, with which the affiliation of a non-located element to A is *doubted*, then $\text{Pl}\,(A)$ is the degree, to which this is *undoubted*, thus taken for understandable or plausible. Naturally, it holds always

$$\text{Pl}\,(A) \geq \text{Cr}\,(A) . \tag{5.34}$$

With respect to further properties and examples of application see BAN-DEMER/GOTTWALD (1995) and the literature cited there, first of all SMETS (1981).

5.2.2 Possibilistic Inference

Whereas in modelling partial ignorance probability measures constituted the starting point of specification and inference, *possibilistic* inference uses only properties of possibility.

A *possibilistic variable* is defined in analogy to the notion of a random variable (see Subsect. 4.1.3). It is given by a *domain* $W \subseteq U$ and a *possibility distribution* $\pi(x)$ for $x \in W$ according to (5.9). As a *realization* of such a possibilistic variable *provisionally* a concrete (crisp) element x_0 of the domain W will be regarded.

A possible idea of the model can be obtained by considering the elements of the domain as individuals, which have partly equal and partly possibly different degrees of possibility of a *breakthrough*, and that at a certain time the breakthrough of such an individual occurs and is observed. In contrast to probabilistic modelling one does not consider here an actual *infinite* period of time, in which all the individuals experience their breakthroughs with each different frequencies.

In contrast to Subsect. 3.2.2 the observed data themselves are specified here *not* as fuzzy sets, but the specification of uncertainty and fuzziness is effected by the modelling of the data genesis, i.e. of their uncertain emergence, or, as one often used to say, of their *environment*.

In practical application this happens, e.g., in taking the given pseudo-exact data, or, if the observation results are given as fuzzy sets, e.g., in restricting the single fuzzy data (provisionally) to their respective (one-dimensional) cores.

In analogy to the sample in probabilistic inference the observation vector $\mathbf{x_0} = (x_{10}, \ldots, x_{n0})$ is assumed to be a realization of a vector of possibilistic variables $\mathbf{X} = (\mathcal{X}_1, \ldots, \mathcal{X}_n)$. The condition of independence of the components of the random sample vector is replaced by the condition that the components should be *minimum-related*: Let be $\pi_i(x_i)$ the possibility distribution of the

i-th component, then the possibility distribution of the observation vector should be given by

$$\pi_{\mathbf{X}} = \min\{\pi_1(x_1), \ldots, \pi_n(x_n)\} \ . \tag{5.35}$$

In analogy to probabilistic inference it is now assumed that the type of the possibility distribution is known up to a (possibly multidimensional) parameter c, which should be *estimated* using the given observation results. Let denote $\pi(x; c)$ the given type of the possibility distribution (the setup) and c the unknown parameter, then every estimation C^* is again a possibilistic variable, which depends on the possibilistic distribution of the observation vector. According to the equivalence of a possibility distribution with the membership function of its inducing fuzzy set (see in (5.10)), the connection rules for fuzzy sets can be used also for possibility distributions. For every possible parameter value c one obtains in such manner via the extension principle and with respect to this value the *conditional possibility distribution* induced by the observation vector and using the supposed minimum-relatedness

$$\pi_{c^*|c}(t) = \sup_{c^*(x_1,\ldots,x_n)=t} \min\{\pi(x_{10}, c), \ldots, \pi(x_{n0}, c)\} \ . \tag{5.36}$$

From here a plausible estimating value c_1^* for c can be computed, which reaches the *highest degree of possibility* according to this distribution:

$$c_1^* = \arg\max_c \min\{\pi(x_{10}, c), \ldots, \pi(x_{n0}, c)\} \ . \tag{5.37}$$

This estimation is called *maximum possibility estimation* and corresponds to the maximum likelihood estimation of probabilistic inference (see some textbook on mathematical statistics with respect to this estimation).

However, one is not forced to use even this estimation, which is very sensitive to outliers. It can happen that the minimum in (5.37) vanishes for all possible c, because the "spread" of the observation values is too large and always a $\pi(x_{i0}, c)$ exists, which equals to zero for even this c. Hence also other estimations can be considered, which are more robust against *outliers*, e.g.

$$c_2^* = \arg\max_c \frac{1}{n} \sum_{i=1}^{n} \pi(x_{i0}, c) \ , \tag{5.38}$$

$$c_3^* = \arg\max_c \max\{\pi(x_{10}, c), \ldots, \pi(x_{n0}, c)\} \ . \tag{5.39}$$

As the estimating procedures of probabilistic inference also the estimating procedures of possibilistic inference can be assessed with respect to their properties within the framework of a decision theory. Remarks to this topic can be found e.g. in BANDEMER/NÄTHER (1992), especially in Chap. 7 there.

A generalization of possibilistic inference to *fuzzy* observations $(\mathcal{A}_1, \ldots, \mathcal{A}_n)$ is now simply possible by another application of the extension principle to the formulae (5.37–5.39) for the crisp case.

A further generalization to the evaluation of functional relationships, in which the multidimensional parameter c to be estimated means, e.g., the coefficients of the functional relationship, is possible in simple manner, although perhaps with much higher computational efforts.

The here presented approach, however, is *absolutely different* from that one in data analysis, treated in Subsect. 7.2.2, in which there is no modelling of the environment, but only the fuzziness of the observation results is transferred into the parameter domain of the relationship. A comparison of these approaches, possibilistic inference and explorative evaluation, must take into account the basically difference in the conditions put and the assumptions used, and is, in general, of no use.

Finally, the important role of *subjective* specifications in handling fuzzy measures should kept always in mind, when using such means to solve practical problems. Hence numerical preciseness does not make sense and the concluded statements should undergo reasonable control by the scientists involved. Nevertheless, fuzzy measures offer powerful means, when the state of knowledge is rather weak.

5.3 Probability and Fuzziness

In the preceding section *fuzzy measures* are introduced for *crisp* sets and should express the degree of possibility, of necessity, of probability, of credibility or of plausibility, respectively, that a *non-located* element is situated or will be situated within a crisp argument set. In practical application, however, such sets can be specified only *fuzzily*, e.g. the set of potential customers for a newly introduced product, the set of locations belonging to a special climatic region, the set of symptoms characterizing a special disease, or the set of worn out components due for replacement. The fuzzy measures mentioned above can now be generalized to this case, i.e., mathematically speaking, they can be defined over the set $\mathbb{F}(U)$, the set of all fuzzy sets over the universe U.

So, for a possibility measure, for which a possibility distribution $\pi(x)$ exists, the defining formula (5.9) can be written in another form using the characteristic function $\chi_A(x) = 1$ for all $x \in A$ and $\chi_A(x) = 0$ for $x \notin A$

$$\text{Poss}\,(A) = \sup_{x \in A} \pi(x) = \sup_{x \in U} \min\{\pi(x), \chi_A(x)\}\,. \tag{5.40}$$

If the characteristic function in this formula is replaced by the membership function of the now fuzzy set $\mathcal{A}$, i.e.

$$\text{Poss}\,(\mathcal{A}) = \sup_{x \in U} \min\{\pi(x), \mu_A(x)\} \tag{5.41}$$

then the desired generalization is obtained.

With respect to a generalization of credibility and plausibility measures it is referred , e.g., to BANDEMER/GOTTWALD (1995).

Because of its importance for application, however, the generalization of probability measures to fuzzy arguments is considered in some detail in the following.

5.3.1 Probability of Fuzzy Events

In a certain analogy to the proceeding with the degree of possibility a continuous probability measure is considered, the distribution of which is based by a density $f(x)$. Then the probability of a *crisp* event can be represented by an integral

$$P(A) = \int_A f(x)\mathrm{d}x = \int_U \chi_A(x)f(x)\mathrm{d}x , \qquad (5.42)$$

from which, replacing the characteristic function by the membership function, immediately the probability of a *fuzzy* event $\mathcal{A}$ is obtained

$$P(\mathcal{A}) = \int_U \mu_A(x)f(x)\mathrm{d}x . \qquad (5.43)$$

The analogous construction for *discrete* probabilities is obvious.

This representation was the starting point taken by ZADEH (1968) for defining probabilities for fuzzy sets. The concept proved workable in many applications, although the usual interpretation of probability as the measure of chance for the event that the next realization will fall into the *crisp* set A rises considerable difficulties in comprehension: the position of a *crisp* singleton, the realization, within the fuzzy set $\mathcal{A}$ would be possible, according to the principle of inclusion for fuzzy sets, only within the *core* A_1 of $\mathcal{A}$. Here an interpretation of $P(\mathcal{A})$ according to (5.43) may help as the *expected value* of the membership function $\mu(X)$, (comp. (4.16))

$$P(\mathcal{A}) = \mathrm{E}_P\mu_A(X) , \qquad (5.44)$$

where the index P should point to the distribution, with respect to which the expected value is to be computed; in the example above this would mean the density $f(x)$.

In this manner the *statistical* interpretation of this probability becomes possible. Let be $x_1, x_2, \ldots, x_n$ independent realizations of a random variable X with the distribution P and X_i the components of the corresponding (mathematical) sample, then on certain little restrictive conditions, one obtains statements like those known from the laws of large numbers in probability theory, (see Subsect. 4.1.4), e.g.

$$P(\mathcal{A}) = \lim_{n \to \infty} \frac{1}{n} \sum_{i=1}^{n} \mu_A(X_i) . \qquad (5.45)$$

For a fuzzy set probability can be interpreted as the *mean membership degree* of the elements of a sample of infinite size from the population according to the probability distributions P.

This probability measure shows some known properties as monotonicity $(\mathcal{A} \subseteq \mathcal{B} \Rightarrow P(\mathcal{A}) \leq P(\mathcal{B}))$ and the formula of generalized addition

$$P(\mathcal{A} \cup \mathcal{B}) = P(\mathcal{A}) + P(\mathcal{B}) - P(\mathcal{A} \cap \mathcal{B}) . \tag{5.46}$$

A transfer of the notion of *independence* of two fuzzy events is *not* possible, as long as intersection is defined by the minimum of the corresponding membership values. If, however, the algebraic product is chosen for this purpose (and correspondingly the algebraic sum for the union) (see (3.30) and (3.31), respectively), then the usual form is yielded

$$\mathcal{A}, \mathcal{B} \quad \text{independent} \quad \Longleftrightarrow \quad P(\mathcal{A} \cdot \mathcal{B}) = P(\mathcal{A}) \cdot P(\mathcal{B}) \tag{5.47}$$

and hence even an approach is opened to define *conditional* probabilities and connections in the sense of BAYESIAN theory.

The introduction of fuzzy sets as arguments of probability takes into account the observation fuzziness, frequently met in practice, and allows even to endow linguistic variables with probabilities and to treat them with methods of mathematical statistics.

Another approach to include observational fuzziness was chosen by KRUSE/ MEYER (1987). As a starting point they consider the case that the realizations of a really *crisp* random variable Y can be observed only *fuzzily*. Such an observation is defined as a realization of a *fuzzy coarsening*, a fuzzy random variable Z, from which the characteristic values of the original Y are to be inferred. For this purpose by $\mathcal{N}$ the set of all possible originals Y is introduced and the fuzzy random variable Z is assumed given. The set of all originals, which can belong to this Z is a fuzzy set over $\mathcal{N}$. If the set of possible originals can be parametrized, e.g. by assuming a certain type of distributions, then fuzziness can be transferred to obtain a fuzzy set over the parameter domain. By an extension principle all formulae used in probability theory can then be fuzzified. This is effected best, as assumed usually, if the realizations are fuzzy numbers or intervals. The already mentioned (see Subsect. 4.2.4) probabilistic inference from fuzzy data in the sense of VIERTL (1996) belongs to this environment. For further details of the KRUSE-approach it is referred to KRUSE/MEYER (1987).

Up to now probability was considered for fuzzy sets. However, even probability can become a fuzzy set, e.g. when stating "the reliability is high", which can be expressed equivalently by "the *probability* of a failure in a time interval of given length is "small".

For the sake of simplicity a universe consisting of *finite* many (N) elements is considered. Then as the probability a vector $\mathbf{P}$ of single probabilities $\mathcal{P}_i(x_i)$ for the elementary probabilities $x_i \in U$ is to be specified. Even if these probabilities are fuzzy, the probability for the certain event, that any element

$x_i \in U$ occurs as an event at all (which corresponds to the universe U itself) must be exact, hence must be *crisply* equal to 1:

$$\sum_{i=1}^{N} P_i(x_i) = 1 \,. \tag{5.48}$$

When computing fuzzy probabilities of events A, which are not elementary events, according to the extension principle this side condition must be taken into account.

The problem becomes essentially more complicated, if the universe is a continuous set, e.g. an interval. Hence with respect to this case it is referred to the book by DUBOIS/PRADE (1980) and to the literature cited in it.

In the considerations on fuzzy probability, up to now, the random events themselves are assumed as *crisp* sets. In some cases, however, it seems to be scarcely comprehensible that fuzzy events are evaluated by crisp probabilities. So, in investigating the ageing process of components one could obtain the statement that the probability of the random event "the component is worn out" equals precisely 0.84. Such an exact statement could be felt a rather artificial one, whereas the statement "the probability is high" would be understood as an appropriate one with respect to the fuzziness of the specification of the whole problem. There ar several suggestions on how to come to reasonable fuzzy probabilities for fuzzy events. Their explicit treatment would be beyond the scope of the present book and its mathematical level. Hence it may suffice to mention the respective basic ideas and to refer to the literature.

Naturally, in turning to fuzzy argument sets starting from fuzzy probability one can choose the way via an extension principle, taking into account the side condition (5.48) mentioned above. This leads, already in the case of more than two elementary events, to the necessity of solving optimization problems (DUBOIS/PRADE (1980)).

Another way was taken by YAGER (1984). He considered the α-cuts of the fuzzy argument sets $\mathcal{A}$. Then the fuzzy probability is introduced as fuzzy set over the *crisp* probabilities $P(A_\alpha)$ as

$$\mathcal{P}(\mathcal{A}) : \mu_{P(A)}(P(A_\alpha)) = \alpha \,. \tag{5.49}$$

YAGER remarks that this can be interpreted as "probability of an at least to the degree α existing satisfaction of the condition A". The so defined probability is denoted by $\mathcal{P}^*(\mathcal{A})$. With $\mathcal{P}_*(\mathcal{A}) = (\mathcal{P}^*(\mathcal{A}^c))^c$ YAGER obtained finally in his cited paper by

$$\mathcal{P}(\mathcal{A}) = \mathcal{P}^*(\mathcal{A}) \cap \mathcal{P}_*(\mathcal{A}) \tag{5.50}$$

the desired probability. With respect to the properties of this fuzzy probability see YAGER (1984).

5.3.2 Random Fuzzy Sets

Finally, one can consider situations, where the fuzzy sets themselves are even *random objects*, e.g. random grey-tone pictures, and hence values of *random fuzzy sets*, as they are called. For an evaluation of the randomly influenced degree Z of components of being worn out one can agree on, e.g., several values of a linguistic variable u over a scale $[0, 100]$: $\mathcal{A}_1 = $ *badly worn out*; $\mathcal{A}_2 = $ *rather worn out*; $\cdots \mathcal{A}_m = $ *useful*. Let the probabilities be known

$$P(u = \mathcal{A}_i) = p_i, \quad i = 1, \dots, m . \tag{5.51}$$

Such problems for discrete (with respect to the finite number of possible values for Z) random fuzzy sets are treated by NAHMIAS (1979). So, the average degree of being worn out is defined as expected value E Z, in explaining the usual weighted sum $\sum_{i=1}^{m} p_i \mathcal{A}_i$ as sum of fuzzy numbers and computing it according to an extension principle.

In the approach due to PURI/RALESCU (1986) the concept of random sets on Euclidian spaces (see for this topic e.g. MATHERON (1975), STOYAN/KENDALL/MECKE (1995)) is fuzzified. For application of this it is interesting that the interpretation of fuzzy data as realizations of random fuzzy sets in the sense of PURI/RALESCU (1986) offers possibilities to transfer methods of mathematical morphology (see SERRA (1982)), as used for crisp sets in image processing, to fuzzy sets, i.e. grey-tone pictures. With respect to erosion and dilation of grey-tone pictures with fuzzy structure elements see GOETCHERIAN (1980) and BANDEMER/KRAUT/NÄTHER (1989). Worth mentioning in this context is the fact that the membership function of a non-random set can be interpreted as the *projection shadow*, as it is called, of a random crisp set (see, e.g., WANG/SANCHEZ (1982)). More precisely formulated, this means that the membership degree $\mu_A(x)$ can be considered as the probability that a corresponding random set S covers the point x (one-point-coverage probability):

$$\mu_A(x) = P(x \in S) = E\mu_S(x) . \tag{5.52}$$

On the other side every random set S is connected with a certain fuzzy set $\mathcal{A}$ (see for details e.g. GOODMAN/NGUYEN (1985)). Recently these mathematical objects are subsumed to fuzzy set-valued random variables (see LI/OGURA/KREINOVICH (2002) with respect to the state of art). This connection between fuzzy and random sets can sometimes be used in situations, in which several kinds of uncertainty are latent and further information gathering would require different experimental expenditure of time, money, equipment, and activated specialized knowledge. Examples of such situations are provided by cases, where really crisp random variables can or will be observed only fuzzily, e.g. because more complicated experiments would exceed the given resources (see for further considerations of such topics BANDEMER/NÄTHER (1992)).

6

Methods from Qualitative Data Analysis

The adjective "qualitative" should point to the character of the statements, which are required in this branch of data analysis. The ultimate aim are not *numerical* results, but a grouping of data according to qualitative criteria of rather different kind. The character of the data themselves (see Subsect. 2.1.1) is unimportant first, although also here the usual starting point are numerical or binary data.

The main aim is a partition of the given data set into subsets (cluster analysis) and the development of procedures to classify further data (classification, discriminance analysis). Every statistical software tool contains procedures to solve these problems, however, only seldom the mathematical background of them is illuminated and the choice of the offered methods is founded.

In this chapter not only hints are given for an assessment of results of cluster analysis, but also some new ideas are presented to get such procedures more flexible and bind them nearer to the context of the given situation.

6.1 Crisp Classification of Crisp Data

6.1.1 The Problem of Cluster Analysis

The problem occurs, as a rule, in preparation of tasks of diagnosis and therapy, of decision making and control.

A given set $O = \{o_1, \ldots, o_N\}$ of distinguishable objects or situations should be partitioned into subsets, into *clusters*, as they are called, in such a manner that the objects *within* such a subset are very similar (*as much as possible*) and that objects from *different* subsets are only little similar to each other (*as less as possible*). This possibility for a cognition of "similarity" is used e.g. in biology to form new species, in medicine for diagnosis and therapy, in technology to detect causes, in general to discover causal relationships.

The fuzzy formulation ("as possible") will later invite for a treatment by methods of the theory of fuzzy sets. In this section, however, the treatment should be effected on the basis of "classical" methods.

The proceeding in partition the given object set into clusters must be, as a rule, highly context depending and heuristic. It includes usually the following three stages (see e.g. ANDERBERG (1973), HARTIGAN (1975), BEZDEK (1981)):

(a) The specification of those *features*, which are assumed to be essential: *feature selection*.

(b) The choice of rules for aggregation and transformation of the different feature shapes observed on the objects and the representation of them as a set of comprehensible mathematical objects (points, functions, graphs, standard forms, etc.) and of rules for decisions on allocation of certain objects to certain subsets: *cluster analysis*.

(c) The confrontation of the result of cluster analysis with the given practical problem for a decision, whether the obtained and suggested partition does make sense in the given situation: *cluster validation*.

Finally, in general, specifications are necessary, how future objects should be allocated to the given clusters: *classification*.

When the partition in clusters is used as a procedure of diagnosis, e.g. by means of a *discriminance function* in the feature domain, then it is spoken of a *discriminance procedure*.

The heuristic of decision making is frequently supported or padded by model ideas, e.g. from mathematical statistics.

For a mathematical treatment the partition task must be represented in a mathematical form:

For every object o_j of the set O then t features F_i are selected, the single shapes of which are specified each by the "value" x_{ij}. In general, the "values" can obviously assume any form of a datum (see Subsect. 2.1.1): cardinal, nominal, quantitative, or set-valued as function, surface or picture, or even as a verbally formulated finding.

The known procedures of crisp cluster analysis, however, require *crisp* items for the feature values x_{ij}, i.e., either real numbers or qualitative yes-no-statements (with "1" for yes and "0" for no).

As can be seen, already by the *selection* of the features to be taken into consideration and their *numerical coding* essential decisions are made influencing the result of the cluster analysis seriously. The *loss of information* suffered in this manner cannot be compensated by any mathematical treatment, however subtle it might be. On the contrary, many procedures of cluster analysis tempt to "information creation" by introducing assumptions that ease the mathematical treatment and decision making, but which are not protected by practical findings.

The usually offered way out to effect the selection of features just in the course of investigation (*feature discrimination*) applies only to the initially already included features and cannot enlarge this subjective selection.

Starting point of a cluster analysis is hence the *data set*, as it is called, or the *data matrix*

$$\mathbf{X} = ((x_{ij})) \,. \tag{6.1}$$

According to this data matrix the N objects o_j should be partitioned into n clusters $C_1, \ldots, C_n$, where n is given and each single object should be allocated to exactly *one* of the clusters.

6.1.2 Mathematical Formulation of the Problem

As already mentioned in presenting the problem, objects are to be allocated according to their *similarity*. For this purpose a *similarity relation* R is to be specified. Over the cartesian product $O \times O$ the (really fuzzy) relation is defined with values on the positive real axis or in the interval $[0, 1]$. It is required that the values should increase with growing similarity and, as a rule, that the relation is symmetric. For the sake of brevity

$$o_j R o_k = s_{jk} \tag{6.2}$$

the *similarity degree* between the objects o_j and o_k is introduced. Then

$$\mathbf{S} = ((s_{jk})) \tag{6.3}$$

is the *similarity matrix* of the object set O.

The specification of this similarity matrix depends strongly on the character of the given data.

In taxonomy, e.g., a branch of biology, only the presence or absence of single features is recorded, to develop a grouping of plants and animals according to their similarity. With such qualitative data (1 if present, 0 if absent) the determination of similarity is effected by the number of correpondences. So, e.g., SNEATH (1957) suggested to divide the number of those features, which are *simultaneously* present at both objects, by the number of features, which are present at both objects *at all* to obtain the corresponding value s.

If the shapes x_{ij} are given be real numbers, then the introduction of a suitably defined distance $d(j, k)$ between the objects o_j and o_k is recommended. In such a way the *distance matrix* of the object set

$$\mathbf{D} = ((d_{jk})) \tag{6.4}$$

is introduced. Because similarity decreases with increasing distance, d is sometimes called also *dissimilarity coefficient*. In every case a transition from a distance d to a similarity relation with values s is possible, even in different ways. One has only to keep in mind that monotonicity inverts its direction and that $s(0) = 1$ and $s(\infty) = 0$ are reached. Then the similarity matrix has the form

$$\mathbf{S} = ((s(d_{jk}))) \,. \tag{6.5}$$

If some features are specified only qualitatively and others quantitatively, then always a common similarity matrix can be constructed, in the simplest case by a weighted sum of the s_{jk} according to (6.2) and the $s(d_{jk})$ according to (6.5).

One should realize that the introduction of such a combination rule will influence the result of cluster analysis essentially. Already the choice of a formula to specify the similarity values s_{jk} and even more of the distance function d for *all features in common* is one of the most critical points of the whole procedure. Even if the range of possible distance functions (see e.g. BANDE-MER/NÄTHER (1992)) and the possibilities of their factorwise combination is enormous, by which the adaption according to the problem should be given, the practical translation is, as a rule, impossible even because of this variety of possibilities. Moreover, there is a tendency to choose the known Euclidean distances, how they are used in mathematical statistics, which is even supported by the common interpretation of the feature values as realizations of random variables (see for this problem area, e.g., MARDIA/KENT/BIBBY (1979)).

To obtain a unique partition of the object set O in clusters $C_1, \ldots, C_n$, a decision rule for the purpose is required, which brings the fuzzy demand: "objects similar as much as possible within each cluster – objects similar as less as possible in different clusters" into a formulation, which can be handled by classical mathematics. As can be seen this is a problem including *two* different criteria.

Let be G a functional, which reflects the similarity of the objects *within* each single cluster $C_1, \ldots, C_n$; it supplies values, which are the smaller the more similar the objects are. Hence one has to demand that

$$G(\mathbf{S}; C_1, \ldots, C_n) = \min_{C_1, \ldots, C_n} , \qquad (6.6)$$

where reasonably $G(\mathbf{S}; o_1, \ldots, o_N) = 0$ and $G(\mathbf{S}; O) = \max$ should be valid. Here the number of desired clusters is left open yet. Moreover, it would be reasonable, if the *union* of clusters would raise the functional value, since the dissimilarity increases in the united cluster:

$$G(\mathbf{S}; C_1 \cup C_2, C_3, \ldots, C_n) \geq G(\mathbf{S}; C_1, \ldots, C_n) . \qquad (6.7)$$

Let be H a functional, which reflects the similarity of objects each of a different cluster; it assumes values, which are the larger the more similar the objects are:

$$H(\mathbf{S}; C_1, \ldots, C_n) = \max_{C_1, \ldots, C_n} , \qquad (6.8)$$

where, correspondingly, $H(\mathbf{S}; o_1, \ldots, o_N) = \max$ and $H(\mathbf{S}; O) = \min$ are to be demanded. Moreover, it will make sense, if

$$H(\mathbf{S}; C_1 \cup C_2, C_3, \ldots, C_n) \leq H(\mathbf{S}; C_1, \ldots, C_n) . \qquad (6.9)$$

Both these criteria must be satisfied by one and the same cluster partition $C_1, \ldots, C_n$ *simultaneously*. Hence a cluster partition is to be computed, for which

$$G(\mathbf{S}; C_1, \ldots, C_n) = \min \qquad (6.10)$$

and

$$H(\mathbf{S}; C_1, \ldots, C_n) = \max . \qquad (6.11)$$

are valid. In certain special cases both these conditions are united, e.g. by

$$G(\mathbf{S}; C_1, \ldots, C_n) - H(\mathbf{S}; C_1, \ldots, C_n) . \qquad (6.12)$$

In every clustering procedure one will recognize these two components.

6.1.3 Some Procedures for Crisp Cluster Partition

Which of the proposals for similarity of objects and for possible cluster partitions should be used depends largely on the practical problem. Strictly scientifically such a decision should be made *after* a "training course"(a *learning period*, as it may be called): objects of known allocation are classified according to factual aspects as well as by different cluster concepts. The results are compared and the decision is made on the expediency of the chosen concepts. Hence in recent times the application of *neural networks* is so popular. By this means, namely, the parameters of corresponding classification procedures, which are specified as neural networks, are computed within a learning period. Though also for this purpose it is necessary to code the single objects by vectors of numbers, all of the same structure. (With respect to neural networks see Subsect. 6.1.5.)

In many concrete cases, however, a preliminary phase of learning would lead either to a unjustifiable expenditure or to not uniquely usable statements. Hence there is an interest in methods, which solve the optimization problem (6.10), (6.11) *in tendency*, where, in general, *dissimilarity* (the functional H) is given priority.

With the strong restriction that the distance between the objects is specified as the Euclidean distance of the feature vectors $\mathbf{x_j} = (x_{1j}, x_{2j}, \ldots, x_{tj})$, i.e.

$$d_{jk}^2 = (\mathbf{x_j} - \mathbf{x_k})^\tau (\mathbf{x_j} - \mathbf{x_k}) = \sum_{m=1}^{t} (x_{mj} - x_{mk})^2 , \qquad (6.13)$$

there exist different proposals for the partition in clusters. (The usually in (6.13) additionally introduced weight matrix $\mathbf{G}$ is omitted for the sake of a simple presentation, it does not change the proceeding methodologically.)

The simplest method consists in the specification of a *similarity threshold* s_0, which corresponds to a distance value d_0. All objects, which have a mutual distance, each of each, that is smaller than d_0, are united to a cluster (SORENSON (1968)).

Usually, however, hierarchical procedure are preferred. Here, initially each single object forms a one-object cluster. Then the two objects with the smallest mutual distance are united to a cluster. In further steps each time those two clusters, the distance of their most distant objects (taken each from the one and from the other cluster) is shortest, are united in *one* cluster. The procedure ends, when *all* objects are united in *one* cluster. In this manner one obtains in the course of the computation all clusters, which could be obtained for arbitrary similarity thresholds according to the preceding proposal.

This cluster partition could be evaluated either subjectively according to factual aspects or by functionals G and H specified beforehand.

The proposals mentioned above can be modified or improved in several directions. So, one can, e.g., choose one of the objects *by chance* and all objects up to a given maximum distance from it unite to a cluster. With the remaining objects the procedure is repeated, and so on (for further details see BONNER (1964)).

The best known of this kind of procedures is that suggested by BALL/HALL (1965) and called ISODATA (described in more detail e.g. in BEZDEK (1981)):

(1) Choose k objects out of O by chance (the "centres" of the clusters);
(2) allocate the remaining $N - k$ objects to these centres, each to the nearest centre;
(3) compute the centres of gravity $\bar{\mathbf{x}}$ for each of the so obtained *preliminary clusters*;
(4) unite those clusters, for which the distance of their centres of gravity is less than a given threshold τ_0;
(5) split up all those single clusters, for which the "inner variation"

$$c_0 \sum_j (\mathbf{x_j} - \bar{\mathbf{x}})^\tau (\mathbf{x_j} - \bar{\mathbf{x}}) \tag{6.14}$$

is larger than a given upper bound; c_0 is the inverse value of the number of objects within the just considered single cluster.
(6) repeat the steps 4 and 5 as often as necessary.

There is a variety of similar procedures.

The main reason for the popularity of the Euclidean distance is the intuitively appealing connection with the variance terms in mathematical statistics. There are different objections to this "minimum-variance" approach to cluster analysis. So, e.g., a change of the scale for only one feature can cause a changing of the whole cluster partition. Therefore it is generally recommended to *normalize* the data, i.e. to centre them to a mean value 0 and divide them by a factor such that the correponding "variance" equals to 1. A clustering method would be desirable, which remains invariant against the numerous families of scale transformations. This demand is largely fulfilled by the fuzzy approach, see Sect. 6.3.

The procedures mentioned up to now pursue the problem of an *optimal* partition only *in tendency*, because they consider only a few possible partitions, the construction of which seems reasonable. If, however, a criterion is realized as suitable for the given situation, e.g.,

$$K(\mathbf{S}; C_1, \ldots, C_n) = K^*(G, H) = \min_{C_1, \ldots, C_n} , \qquad (6.15)$$

then the solution task arises for this optimization problem. It is a *combinatorial* problem and hence always solvable *theoretically*, because all sets involved are *finite*. The method for the solution is called *total enumeration* and consists in a listing of all possible constellations. Naturally, the number of cases to be considered is enormous already for small numbers of objects. For example for 8 objects and 3 possible clusters already 966 values of K^* are to be computed. Hence this procedure is practicable only for such situations, where the relations can be looked over with naked eye and knowledge of the subject. Therefore iterative procedures from dynamical and integer programming are used to come to acceptable results for realistic numbers of objects essentially more rapidly. Obviously, all this expenditure is worthwhile, however, only if both, the chosen types of distance as well as the optimization criterion have a solid basis by specialized knowledge on the situation.

For hierarchic procedures, which produce series of sets C^s of clusters, e.g. $C^1, C^2, \ldots, C^r = O$, the results of the partitions can be presented very clearly in graphical form, if the number N of objects is not too large. Figure 6.1 shows an example of such a *dendrogram* by which can be seen, how the clusters were united with decreasing similarity. Obviously, the uniting itself could be effected also according to another criterion.

Specially, if the objects are allocated according to the principle of nearest neighbourhood, then there is a connection to graph theory; in this case

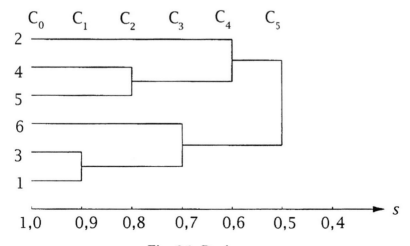

Fig. 6.1. Dendrogram

Table 6.1. shows a possible similarity matrix **S** corresponding to the dendrogram in Fig. 6.1

$$
\begin{pmatrix}
1 & 0.5 & 0.9 & 0.5 & 0.5 & 0.7 \\
0.5 & 1 & 0.5 & 0.6 & 0.6 & 0.5 \\
0.9 & 0.5 & 1 & 0.5 & 0.5 & 0.7 \\
0.5 & 0.6 & 0.5 & 1 & 0.8 & 0.5 \\
0.5 & 0.6 & 0.5 & 0.6 & 1 & 0.5 \\
0.7 & 0.5 & 0.7 & 0.5 & 0.5 & 1
\end{pmatrix}
$$

the dendrogram is represented as a *tree*, as it is called. (For details see, e.g. JARDINE/SIBSON (1971).)

6.1.4 Clustering with a Mathematical-Statistical Background

Naturally, the solution of the partition problem becomes clearer, if there are assumptions on the *genesis* of the data, here the feature values, which are based on technical considerations or scientific plausibility. The usual assumption that these data are realizations of random variables are only seldom scientifically founded, even as a rule, when software is used, it is hardly questioned or made transparent.

Most clearly the relation to probability theory can be seen in *modal analysis* after WISHART (1969). For the sake of simple presentation the case of only *one* feature is considered. It is assumed that the number of objects are sufficiently high such that the histogram of the feature values observed is already meaningful, in the sense of mathematical statistics. The procedure investigates at first, whether the probability distribution basing the data is multi-modal, i.e. whether its density shows several maxima. From this fact it is concluded that the distribution is a *mixture* of several distributions, which sufficiently contrast from each other. The aim of the partition means here the allocation of each single object to one of these different distributions. The procedure consists in that at first the histogram of the feature values is established, in which the respective relation of each value to its object can be seen. Then the objects are allocated to different regions with high frequencies (the modal regions of the histogram) and the intermediate regions with low frequencies are left out of consideration at first. Finally, the remaining objects are allocated each to that cluster, to which their respective feature values are nearest. In the case of several features this method is, as a rule, impractical.

Hence WISHART (1969) suggested a threshold-algorithm to recognize regions with high frequencies as initial clusters. With a given frequency threshold w_0 and a distance threshold d_0 for each object o_j the frequency w_j of all objects with a smaller distance than d_0 is computed:

$$
w_j = \text{number } \{k \mid d(j,k) \leq d_0\} . \tag{6.16}
$$

Then, at first, all those objects are left out of consideration, for which it holds $w_j < w_0$. The remaining objects are allocated to clusters by gathering suitable nearest neighbours. Then those objects left out of consideration are allocated to these clusters according to any useful criterion. This is only *one* possible procedure that should show the general proceeding. Also the ISODATA-procedure is such a procedure, if the feature values are assumed to be realizations of random variables.

The problem changes again, if one can assume that the n different distributions, which form the mixture for the distribution of the feature values, are all *known* exactly. What matter is then, obviously, to allocate each single object to its respective distribution correctly. This allocation specifies then the different clusters. In this case the starting point and the criterion for the cluster partition of all objects consists in the *likelihood function* for all feature values, interpreted as a sample. The allocation of the objects to the clusters is then effected according to the reasonable principle of regarding that special distribution as the true one (and allocating the object to the respective cluster), for which the likelihood function of the feature value is largest. If the distributions are all normal, then (the logarithm of) the likelihood function is a square function, one comes back to square distances, how they are already considered with the heuristics in the preceding subsections (See for details and this case the book by MARDIA/KENT/BIBBY (1979).)

The *classification*, i.e. the allocation of *further* objects to the already obtained clusters, when should be considered as a routine task with a statistical background, is called *discriminance analysis*. The procedures developed and founded for this case within the framework of mathematical statistics, are used in practise, passing over in silence, also in such cases, in which this background is not given at all; then they remain heuristics, naturally, as all other procedures mentioned in the preceding subsection.

The background model has the following form: It is assumed that the vector of the feature shapes $\mathbf{x}$ of an object o_j is a realization of a random vector $\mathbf{X}$, the density of which is one of n possible densities $f_i(\mathbf{x})$. The aim of discriminance analysis consists in the allocation of a *given* realization $\mathbf{x}_0$ of a further object o_0 to one of the n populations. It is the idea that the populations had been deduced from the clusters beforehand in some reasonable manner. The *decision* on the allocation is to be effected *as well as possible*, that is interpreted in mathematical statistics that *wrong* decisions should happen as seldom as possible. Hence a *discrimination rule* is a decision rule, which brings about a partition of the sample space (the set of all possible sample vectors) into regions R_i together with the instruction: If the given realization $\mathbf{x}_0$ lies in R_i, then assign the corresponding object o_0 to the population with the density f_i. The task of mathematical statistics consists then in the determination of such regions R_i in dependence on the given densities f_i, on the condition that the probability of a *wrong assignment* is as small as possible.

Naturally, the assumption that the densities are known *exactly* is only little realistic. Hence usually the case is considered that these densities (e.g.

via suitable parameters) are estimated from the clusters determined beforehand. This is one of the methods mentioned above for a connection of cluster problems with those from discriminance analysis.

Also here a BAYESIAN approach exists, in which there are given a-priori probabilities for the assignment to the populations. A typical example from medicine is the assignment of new patients ("objects") from their symptoms ("features") to different diseases ("populations"). The a-priori probabilities would be obtained from the frequencies of the occurrence of these diseases. The assignment follows then according to the largest a-posteriori probability among the diseases given the feature vector $\mathbf{x}$. The practical application of this approach is frequently problematic, e.g. with respect to the origin and the reach of validity of the a-priori probabilities as well as in the case of multimorbidity.

The properties of the different rules of discriminance are investigated usually by Monte-Carlo-simulations or by application to extensive sets of real data, the correct assignment of which is known. If the parameters of the populations are only *estimated*, then an estimation of the probability of wrong classifications turns out, as a rule, to be too optimistic.

A sound scepticism is always reasonable also in the use of standard procedures of cluster and discriminance analysis. However, this should not lead to a general refusal of such procedures as being unsuitable, they remain a reasonable support in problems to make necessary partitions of practical data sets clearly visible and comprehensible.

6.1.5 Basic Ideas of Neural Networks

For the assignment of objects to certain sets the concept of *neural networks* is used already for years. A first instructive presentation of the basic ideas, from the point of view of *artificial intelligence*, is contained in the collection by RUMELHART (1986), with respect to a comprehensible introduction into the theory see e.g. ROJAS (1993) and KOSKO (1992). Starting point of this approach are imaginations on the way, how human thinking operates "technically", which is superior to high-speed computers in some respects. The brain consists, as is well known, in a highly complicated net of *neurons*, which can activate and deactivate each other. This makes possible a high *paralellism* of single events as well as a strong *robustness* against faults and failures in certain parts. The work of the neurons is, first of all, important for *intelligent* concluding, the basis of which consists in recognition and assignment of situations.

The theory of neural networks tries to model, mathematically, this way of action of the net of neurons within the brain and to implement it on suitable computers in a simplified manner. At present the set of concepts can hardly be overlooked. Hence, in the following, only certain simple basic ideas are presented as starting point for some hints for an assessment and use of corresponding brain tools.

First of all, a practically rather uninteresting information situation will be considered: n objects o_j are characterized each by a (column) vector $\mathbf{v}_j$ of n *exact* feature values. As an aim it is demanded that the objects are to be *recognized* in their respective occurrence in a manner *as simple as possible*. Moreover, it is assumed that the n vectors are linearly independent in the corresponding n-dimensional Euclidean space $I\!R^n$ (i.e. that no one of them can be represented by a linear combination of the others). The solution of this task is given by a square matrix

$$\mathbf{W} = ((w_{kl})) , \tag{6.17}$$

which maps the n *input vectors* $\mathbf{v}_j$ on n corresponding *output vectors* $\mathbf{u}_j$

$$\mathbf{u}_j = \mathbf{W}\mathbf{v}_j . \tag{6.18}$$

These output vectors $\mathbf{u}_j$ could be, e.g., the n unit vectors, which have only in the respective jth coordinate the number 1 and zeros in the other ones. In this way a simple distinction of the n possible objects would be guaranteed.

The elements of the weight matrix $\mathbf{W}$ are determined in an easy algebraic manner from the given input and output vectors $\mathbf{v}_j$ and $\mathbf{u}_j$. The determination is, particularly, very easy, if the input vectors are *normalized* and *orthogonal*, i.e. if

$$\mathbf{v}_i^T \mathbf{v}_j = \begin{cases} 0 & \text{for} \quad i \neq j , \\ 1 & \text{for} \quad i = j . \end{cases} \tag{6.19}$$

In this case the network matrix $\mathbf{W}$

$$\mathbf{W} = \sum_{i=1}^{n} \mathbf{u}_i \mathbf{v}_i^T , \tag{6.20}$$

is thus the sum over the matrix products of the connected input-output vector-couples.

In interpreting as a neural network the coordinates of the vectors are interpreted as *neurons* and their present values as their *activations*; the elements of the weight matrix w_{kl} evaluate then the respective *connectivity* between the k-th coordinate of the input and the l-th one of the output according to (6.17).

The information situation becomes immediately interesting *practically*, if the input vectors are allowed to *vary*. Then it concerns feature vectors of objects, for which the allocation only to a *category* can be effected, a problem, which was considered, in a slightly changed expression, already in the clustering problem.

If the ranges of variability of the input vectors are (approximately) known, then one succeeds mostly by specifying of a function for an additional modification of the coordinate values of the computed output vector to come to unique and, as a rule, *correct* allocations.

Thus, e.g., the *threshold function* is such a popular *activating function*, as it is called. In this manner, the value of each coordinate of the output vector, which lies usually between 0 and 1, is put to 0, if a given threshold value is not reached. The threshold values can be chosen differently for different coordinates. Another type of activating functions are the *sigmoid functions s*, which change the original output value u of a coordinate monotonically into a value $g(u) \in [0,1]$, e.g., by

$$s(u) = \frac{1}{(1 + \exp\{-cu\})}; \quad c > 0 \,. \tag{6.21}$$

With such an activating function one succeeds e.g. to suppress a certain basic noise, i.e. slight variations of the input vectors, with respect to the allocation of the objects. One should note that the problem does *not* consist in a *discovering* of a pattern (pattern cognition), but in a *recognizing* (pattern recognition) of a pattern (one of the categories of the inputs).

If the variation of the inputs obeys *stochastical rules*, then these must be known in order to construct the weight matrix and to compute the outputs in such a manner that in spite of these variations an allocation, as correct as possible, on the basis of these outputs can be guaranteed. Moreover, sometimes even a temporal change of these rules can be taken into account, if this change is known, e.g. if the probabilities $p_j(t)$ for the appearance of the representatives of the different categories as functions of time t are given. This would be the case, e.g., when certain appearing species of animals, the feature vectors of which are observed as inputs, are known to be active only at night. Mostly, however, the probabilities of appearance are assumed to be independent of time.

A-priori knowledge of the probability of occurrence of objects, which should be allocated, by neural networks, to categories, are indispensable first of all, if the corresponding input vectors are allowed to be *mutilated*, i.e. if the respective items can be missing for some features. In this case the given probabilities for the occurrence of the n categories considered are interpreted as an *a-priori distribution* and over the mutilated coordinates a probability distribution of maximum entropy is spread out. With respect to SHANNON entropy see some textbook and (6.63).

A possibility to further flexibility of neural networks is given by the introduction of *hidden units*, as they are called. These are neurons (units), which do not appear as supports of values neither of input coordinates nor of output ones. In the simplest case they form a *second layer* after the input and before the output. Some coordinates of the input are mapped, at first, onto this "interim output", which influences for its part the whole network. This interim mapping is represented e.g. by a further matrix. By introduction of *several* interim layers of hidden units the overall configuration of the neural network can be simplified and the number of identifiable categories can be increased essentially. Also in this case the connectivity remains *linear*, because

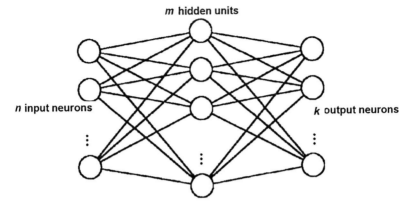

Fig. 6.2. Example of a neural network with a layer of hidden units

the overall matrix of the neural network, which represents the connectivity of input and output, is obtained by matrix multiplication of the single matrices.

If, however, also the interim outputs are subjected to activating functions, then the system becomes really *nonlinear*. Nevertheless, such neural networks have each a domain, in which they behave approximately linearly; hence one finds for this case sometimes the notation *semilinear*.

The attractivity of neural networks is supported essentially by the possibility to determine the elements of the network matrices and of the parameters of the activating functions (e.g. the threshold values or the parameters of the sigmoid functions) sequentially and approximately within a *learning process*.

The learning process starts with adjusting of the required quantities, the elements of the matrices and the parameters, to certain values. Then input-output vector-couples belonging to each other are offered to the neural network in random selection and order and the initial values of the required quantities are changed in a such manner that each time the difference between the given and the just obtained output is as small as possible. For demonstration the simplest case is considered that only one network matrix $\mathbf{W}$ is to be determined and no activating function is used.

Let be given m vector couples $(\mathbf{v}^j, \mathbf{u}^j)$, the connectivity of which is to be learned by the neural network. As the initial matrix $\mathbf{W}(0)$ the zero matrix $\mathbf{O}$ is chosen. Moreover, let be $(\mathbf{v}_i, \mathbf{u}_i)$ the vector-couple offered in the i-th step, then the *delta rule*, as it is called, for the computation of the network matrix $\mathbf{W}(i+1)$ in the next step has the form

$$\mathbf{W}(i+1) = \mathbf{W}(i) + \eta\delta(i)\mathbf{v}_i^T . \tag{6.22}$$

Here $\eta \in (0,1)$ is a given scalar constant, which determines the *learning rate*, as it is called, and $\delta(i)$ is the difference vector between the corresponding and the just computed output vector

$$\delta(i) = \mathbf{u}_i - \mathbf{W}(i)\mathbf{v}_i . \tag{6.23}$$

The value of this learning rule consists in that it can be applied also if the input vectors are chosen no longer as being orthogonal to each other, but that they are only linearly independent. Even if the input vectors are disturbed by realizations of slight random variables, one obtains still useful results. Also in the case that the input vectors be *no longer* linearly independent (or cannot be so because the dimension n of the vectors is smaller than their number m) a network matrix is created, which yields an adaption in the sense of the method of least squares.

A neural network can be interpreted, generally, also as a representation of a function of several variables. Starting with m couples of argument vectors $\mathbf{x}_j, y_j$ a function $y = f(\mathbf{x})$ is to be "learned', which assigns the training output values y_j to the training input values $\mathbf{x}_j$ as precisely as possible (in the sense of the method of least squares). As a special case, the deviations for the training set are all zero and the approximation is an interpolation. Frequently the result obtained by the network is a piecewise constant function (see ROJAS (1993) or some other textbook on this topic). On the other side, by a suitable network a procedure can be constructed, which (in the arguments) corresponds to linear regression and hence yields the same computing formula for the parameter vector. For the setup

$$y = b_0 + \sum_{i=1}^{n} b_i x_i \tag{6.24}$$

one obtains with the design matrix $\mathbf{X}$, the rows of which consist of a 1 (with respect to the constant b_0) and of the vector $\mathbf{x}_j^\tau$, the estimation $\hat{\mathbf{b}}$ of the column vector formed by the coefficients b_i

$$\hat{\mathbf{b}} = (\mathbf{X}^\tau \mathbf{X})^{-1} \mathbf{X}^\tau \mathbf{y} . \tag{6.25}$$

See Subsect. 7.1.2.

By networks with several layers and by use of sigmoids as activating functions even nonlinear connectivities can be represented approximately (according to the criterion of least squares), e.g. with setups of logistic regression

$$y = \frac{1}{1 + \exp\{- \sum_{i=1}^{n} b_i x_i\}} . \tag{6.26}$$

Learning procedures, which utilize minimization of the deviations of the just obtained outputs from the given ones, are called frequently backpropagation algorithms (see e.g. ROJAS (1993)). Backpropagation in networks with several layers is a very popular method to solve practical problems. It is applied e.g. in robotics, in recognition problems for patterns and speach, and for encoding problems. In all these fields of application the aim is in common that with only a few empirical data a network should be designed, by which the basing connectivity can be simulated in a certain sense. In many cases the empirical connectivities of the model variables are unknown from the outset or can change in the course of time. The users are convinced that by back-propagation statistical regularities can be discovered and exploited. Moreover, changes in time can be taken into account by adaption of the network parameters (by permanent learning).

Backpropagation in multi-layer networks is only one direction in the development of theory and practice of neural networks. Similar fields of application, as picture processing, cluster and discriminance analysis, are mentioned, and in each field successful applications are referred to. This led in wide sections of scientists, especially among users, to the euphorical assessment that with the concept of neural networks a *factotum* (literally: a man who does all kinds of work) had been found, by which, in a uniform and natural manner, *every* problem of application can be solved. However, the problems in handling numerical tasks of interpolation, approximation, and discrimination are *not* removed by their couching in the language of neural networks, but only covered up and left out of consideration. This is the *main* danger in general application of neural networks for the using practician: The necessary *critical* point of view on the respective procedure, the input vectors as well as the results obtained, is replaced by fascination and firm belief in their quality.

In this context it can even be left out of consideration that many procedures effected by neural networks turn out to be well-known already for a long time. So, the computation rule for Hopfield networks corresponds to the relaxation method of Southwell, introduced into physics many years ago (see e.g. COURANT/FRIEDRICHS/LEWY (1928); and ROJAS (1993)). Similar remarks can be made also with some recommended cluster and approximation procedures.

Already rather early there were critical considerations on problems of neural networks (see MINSKY/PAPERT (1969)), but today it seems to be rather difficult to hear critical voices out of the mass choir of publications on neural networks.

The remarks made in this book with respect to the treatment of individual practical problems remain valid *also* for neural networks.

Thus already the specification of data for the input as vectors of binary or real numbers is a source of uncertainty and arbitrariness.

For the learning process with "disturbed" inputs, as a rule, there are explicitly no assumptions put on the kind and possible regularities of the disturbance, although these would be important for an assessment of the performance of the network *after* the training period.

Finally, as a rule, the user is unable to follow the way of uncertainty of the input through the network and to assess the numerical behaviour within the network.

If training and working periods should alternate regularly, which is the case frequently in practical application, then the "convergence" of the learning process is assessed by criteria (e.g. by the deviations of training runs following each other), which are not unproblematic even in common sequential processes.

Hence one should combine also neural networks with an accompanying critial point of view. The quality of data used should be proved and the behaviour of the neural network should be investigated by testing with data from problems with simple and known results.

Also a neural network *cannot create* information, one can obtain appropriate results only if the network is offered the necessary information for these, and if the network reflects the practical situation sufficiently precisely.

Finally, a development will be mentioned, in which the concept of networks is combined with that of fuzzy sets: the *fuzzy neural networks*. The fuzzy inputs $\mathcal{V}_i$ are connected with each other according to their character, e.g. by

$$\mathcal{V}_{ij} = \mathcal{V}_i \cap \mathcal{V}_j , \qquad (6.27)$$

which means for the corresponding membership functions $\mu_i(x), \mu_j(x)$ e.g.

$$\mu_{ij}(x) = \min\{\mu_i(x), \mu_j(x)\} . \qquad (6.28)$$

These can be suitably combined to a fuzzy output, e.g.

$$\mathcal{U}_i = \bigcup_j \alpha_{ij} \mathcal{V}_{ij} , \qquad (6.29)$$

where the $\alpha_{ij} \in (0, 1]$ are appropriate parameters with $\sum_j \alpha_{ij} = 1$ for all i.

The just mentioned example is a purely theoretical one, but it can show how the connection of the crisp input values by multiplication and summation, e.g. in (6.18), can be replaced by connection rules for fuzzy sets. As a nice example for the information processing in a fuzzy neural network the colour sense in the human eye is mentioned by ROJAS (1993), which gets by on fuzzy receptors for blue, green, and red, and produces, nevertheless, a picture in the brain with a lot of nuances of colours. In modelling this process it could be shown that functions of a form like density functions of normal distributions ("Gauss' bell curves") are suitable as the membership functions of the three basic colours. In the case that also *mutilated* data are allowed, the probability approach mentioned above can be modified for fuzzy data. Then the probability distribution for the occurrence of the n different categories (e.g. the 10 digits in handwritten and mutilated items in the input) is replaced by a *possibility distribution* $\pi_i; i = 1, ..., n$, and over the mutilated "coordinates" a conditional possibility distribution of maximum entropy is spread out. The entropy of a fuzzy set is given by an entropy measure e, as it is called, e.g. by

$$e(\mathcal{A}) = \frac{\text{card}\,(\mathcal{A} \cap \mathcal{A}^c)}{\text{card}\,(\mathcal{A} \cup \mathcal{A}^c)} . \qquad (6.30)$$

With respect to a detailed representation see e.g. KOSKO (1992). and the journal IEEE Transactions of Neural Networks.

Naturally, the general remarks with respect to the handling of fuzzy sets remain valid also for fuzzy neural networks.

6.2 Fuzzy Classification of Crisp Data

6.2.1 Fuzzy Cluster

In partition of a set of objects O in clusters $C_1, \ldots, C_n$ it is assumed without saying that each object o_j should be allocated to its "correct" cluster. In discriminance analysis this requirement is contained in the problem even explicitly. However, there are practical situations, in which this assumption of uniqueness is no longer realistic, e.g. if a patient has several diseases from the given list of possibilities (multi-morbidity), or if an asked electorate in the run-up to the elections is yet attracted by several parties or if he has several votes, which could be splitted. In many cases, however, it seems rather difficult to allocate an object to one of the clusters, which are felt to be suitable, only from the feature values, because there would be, obviously, too much arbitrariness involved. Even this necessarily subjective assessment, which may have definitely its factual reasons, makes it desirable to allocate those objects to *several* clusters or to each single of these clusters *gradually*. This leads frequently, first of all for feature values in certain "border regions" to a more meaningful interpretation of the partition of the set of objects. Such clusters with gradual memberships of objects are called, consequently, *fuzzy clusters*. With respect to the following considerations in this Sect. 6.2 the feature vectors remain but *crisp* or common vectors of real numbers.

Starting point of the considerations is, as in Sect. 6.1, the similarity matrix $\mathbf{S}$ or the distance matrix

$$\mathbf{D} = ((d_{jk})) \tag{6.31}$$

(compare (6.4)). The allocation of the objects to *crisp* clusters can be represented by a *partition matrix* $\mathbf{Z}$, the elements of which z_{ij} indicate, whether the object o_j belongs to the cluster C_i ($z_{ij} = 1$) or not ($z_{ij} = 0$):

$$\mathbf{Z} = ((z_{ij})) . \tag{6.32}$$

Since in the *crisp* case each object should belong to exactly one cluster, the matrix $\mathbf{Z}$ has the property that for all j

$$\sum_{i=1}^{n} z_{ij} = 1 \tag{6.33}$$

is valid, and because one wants to have at least one cluster that contains at least two objects (in the other case clustering would be senseless) one obtains the condition

$$0 < \sum_{j=1}^{N} z_{ij} < N . \tag{6.34}$$

BEZDEK (1981) called such a partition a *hard n-partition*.

The problem of such a *hard* partition leads, as a rule, to combinatorial tasks, which might demand a high numerical expenditure, as already mentioned with the method of total enumeration. The transition to fuzzy clusters yields therefore additionally also advantages in the numerical treatment of the tasks.

If gradual membership of objects to several clusters is allowed, then both the preceding conditions can be really omitted. However, in order to preserve the *crisp* case as a particular one, BEZDEK retains the condition (6.33). Hence each object o_j must distribute its membership to the n cluster. The part apportioned to the cluster C_i is then $\mu_j(i)$, and it should hold

$$\sum_{i=1}^{n} \mu_j(i) = 1 . \tag{6.35}$$

For a numerical handling it is favourable (BEZDEK (1981)) to consider the case with the condition (6.34) retained (called *fuzzy n-partition*) as well as the case that this condition is moderated to

$$0 \le \sum_{j=1}^{N} \mu_i(j) \le N \tag{6.36}$$

for all i. In this latter case also procedures are conceivable that the procedure itself searches for a favourable number of clusters, because also *empty* clusters are allowed. Such partitions are called *degenerated* fuzzy n-partitions. Moreover, this closure eases the theoretical treatment (see for details the book by BEZDEK (1981) just mentioned).

Obviously, a fuzzy cluster C_i with the membership function $\mu_i(j)$ can be interpreted also as a fuzzy set over the set of objects O. There are different possibilities for what should be understood as *similarity* of such fuzzy sets. This problem is treated within Sect. 6.3 from a point of view somewhat more general.

6.2.2 Procedures of Fuzzy Cluster Analysis

Naturally, every procedure of cluster generation with crisp feature values can be generalized to the case that fuzzy clusters are allowed, e.g. by allocating such objects with approximately equal affinity (similarity, distance) to *several* clusters with appropriate single degrees.

For BEZDEK (1981) first of all such procedures are interesting for a generalization, which use a *Euclidian distance* d_{jk} of the objects and a criterion of optimality (objective function) (see Subsect. 6.1.3). Already RUSPINI (1972) originated the idea to choose the *local density* of the feature vectors as a measure of "quality" of the clustering. He considered the following criterion of optimality

$$J_R(\mathbf{M}) = \sum_{j=1}^{N}\sum_{k=1}^{N}\left\{\left[\sum_{i=1}^{n}\sigma(\mu_i(j)-\mu_i(k))^2\right]-d_{jk}\right\}^2 = \min_{\mu} \qquad (6.37)$$

with the real positive constant σ and a fixed chosen n. The matrix $\mathbf{M}$ consists of the membership values $\mu_i(j)$. Obviously, the value of this functional J_R is small if the single terms of the sum are each small. This happens if couples of objects lying near to each other (i.e. d_{jk} is small) obtain approximately equal membership values to the respective clusters (the difference of the respective μ_i is small). The factor σ makes sure that distances and membership values are comparable in magnitude.

For a determination of an optimal membership matrix $\mathbf{M} = ((\mu_i(j)))$ according to the criterion (6.37) gradient methods are used. Because these methods, naturally, find only *stationary points* in the corresponding space of possible membership values, the obtained recommended clusters should be, absolutely, tested for their meaningfulness (cluster validation).

The original algorithm due to RUSPINI (1972), which is not presented here, is said to be rather difficultly implementable. Its computional efficiency should be weak and its generalization to more than two clusters should be of little success. But it was the pioneer for a successful development of this approach, which was mentioned here for its exemplarity and intuitive comprehensibility of the criterion used.

Hence BEZDEK starts with the case that for each cluster C_i a *virtual* object v_i (cluster centre, prototype) can be defined, the feature values of which are determined as the centres of gravity of the feature values of the corresponding objects. Let denote $\mathbf{x}_j$ the feature vector of the object o_j and $\mathbf{v}_i$ the feature vector of the virtual object v_i, determined by all those objects in the cluster C_i, and let be $d_j(i)$ the Euclidean distance of the object o_j from v_i. Then BEZDEK (1981) had chosen the following functional (fuzzy q-mean-functional):

$$J_q(\mathbf{M}, \mathbf{V}) = \sum_{j=1}^{N}\sum_{i=1}^{n}\left(\mu_i(j)\right)^q d_j^2(i) ; \qquad (6.38)$$

with the matrix $\mathbf{V}$ consisting of the feature vectors of the cluster centres. The parameter $q \in (1, \infty)$ is an arbitrary but given weighting exponent, by which the fuzziness of the clusters can be controlled. The larger the chosen q *the more fuzzy* the assignment of membership to the clusters will be. If q tends to 1, then the clusters form a *hard* partition. For a solution of this special optimization problem BEZDEK suggested an iteration procedure, which should be presented here for the simplest case. With respect to possible generalizations it is referred to the book by BEZDEK and for a short survey to BANDEMER/NÄTHER (1992).

For starting the procedure, generally, one has to choose a Euclidean norm $d_j(i)$, a number n for the number of clusters allowed at the most, a number $q \in (1, \infty)$ and, for a stopping rule, a measure of dissimilarity of matrices $\mathbf{M}$, e.g. a matrix norm, as it is called, and for that a small number (ϵ).

As a starting point one chooses then a matrix $\mathbf{M}^{(0)}$, which has all the properties of a non-degenerated partition matrix (e.g. as a result of a rough crisp clustering).

For $l = 0, 1, 2, \ldots$ one executes the following steps:

(1) Compute the n cluster centres $\{\mathbf{v}_i^{(l)}\}$ according to

$$\mathbf{v}_i = \frac{\sum_{j=1}^{N}(\mu_i(j))^q \mathbf{x}_j}{\sum_{j=1}^{N}(\mu_i(j))^q} \tag{6.39}$$

with $\mathbf{M} = \mathbf{M}^{(0)}$.

(2) Determine a matrix $\mathbf{M}^{(l)}$ using these $\{\mathbf{v}_i^{(l)}\}$:
in defining first the index set of all those $\mathbf{x}_j$, which coincide with certain of the computed centres $\mathbf{v}_i^{(l)}$:

$$I_j^{(l)} = \left\{ i \in \{1, .., n\} : d_j(i)^{(l)} = d(\mathbf{v}_i^{(l)}, \mathbf{x}_j) = 0 \right\}$$

and its complementary set

$$I_j^{(-l)} = \{1, \ldots, n\} \setminus I_j^{(l)} .$$

If $I_j^{(l)}$ is empty, then determine as the new membership values for o_j

$$\mu_i^{(l+1)}(j) = \left[\sum_{s=1}^{n} \left(\frac{d_j^{(l)}(i)}{d_j^{(l)}(s)} \right)^{\frac{2}{(q-1)}} \right]^{-1} . \tag{6.40}$$

If, however, $I_j^{(l)}$ contains at least one index, then put all $\mu_i^{(l+1)}(j)$ for the indices $i \in I_j^{(-l)}$ to 0 and distribute the membership 1 to the clusters with indices from $I_j^{(l)}$.

Both these steps are to be repeated until the "difference" between the matrices $\mathbf{M}^{(l)}, \mathbf{M}^{(l+1)}$ following each other is smaller than the initially chosen ϵ.

Naturally, also the result of this proceeding is only a *suggestion* on how the object set can be allocated to subsets.

Although fuzzy cluster analysis represents a certain progress compared with crisp analysis, because it can react more flexibly to the facts of practical problems, however, an essential restriction remains: the feature values, at last, must be numbers, and the observation fuzziness of these feature values is left out of consideration.

Naturally, methods of cluster analysis supply reasonable results in many practical situations, first of all, if they are controlled by basic knowledge of the applying scientist. There is, however, a whole string of principal objections against cluster analysis, especially if it should go along with a criterion of optimality. It is not only the arbitrariness of such a criterion and its structure, as e.g. the notion of the basing distance and its mathematical formulation and the parameters of the procedure, it is also the vagueness of the notion of optimality itself and, first of all, the necessity to form a *common feature vector* **x** out of partly totally incomparable features, a vector, the similarities or distances of its components determine the result essentially.

Therefore it seems to be important to make oneself aware of the ultimate aim of cluster analysis. Frequently the search for a structure, here for subsets of objects, which are very similar to each other, is only an intermediate target for the problem to allocate further objects to classes (e.g. typical situations in controlling or in diagnosis in medicine) or to find relations among features. Moreover, if the number of features increases, then the problems in carrying out cluster analysis become more and more serious. Hence in such cases methods of local concluding become more and more interesting, by which problems of classification and specification of relations among features can be solved *without* a preliminary total partition of the whole object set by an algorithm of clustering. Moreover, some of the specified features should be allowed to assume *fuzzy data* as values. In this manner, inter alia, also the case of aggregation of fuzzy descriptions of a situation and that of determination of approximate control instructions, are examined from another point of view and solved. With such methods of local concluding in data and knowledge bases the next section will be concerned.

6.3 Fuzzy Classification of Fuzzy Data

The frequently unrealistic assumption that the feature shapes of the single objects can be represented sufficiently precisely by *numbers* is omitted in this section. Features are allowed for consideration, the shapes of which can be described by *fuzzy sets*. In this context the problem of finding a *distance*, in which *all* the features of the objects are included, is not tackled *directly*. The similarity concept will be proved to be a more appropriate way of looking at the problem. That the result of the considerations can also be comprehended *formally* as a specification of a distance again will turn out to be unimportant for practical application.

6.3.1 Fuzzy Similarity of Fuzzy Data

Starting point of cluster analysis was the demand that the objects *within* a cluster should be "as similar as possible" and that objects in *different clusters* should be "as dissimilar as possible". This typically fuzzy formulation was precised in Sect. 6.1 mathematically by introduction of a fuzzy similarity relation R, which provides for each two elements of a universe U the degree on the scale $[0, 1]$, by which they are similar to each other. For points as feature values this was effected by a *distance* d, in which all the different features of an object must be included.

Now a further fuzzy structure will be added. The shapes of the features for the single objects are allowed to be fuzzy sets $\mathcal{X}_{ij}$ each over the respective single feature universe U_i. In this case the data matrix (6.1) has the form

$$\mathbf{X} = ((\mathcal{X}_{ij})) \, . \tag{6.41}$$

As an example of such an $\mathcal{X}_{ij}$ a chemical compund o_j is considered and as a feature F_i its toxicity for the human respiratory tracts. The $\mathcal{X}_{ij}$ would be, e.g., the value *high* of the linguistic variable RESPIRATORY TOXICITY.

For a better comprehension of the contents and for the sake of simplification of the mathematical representation in this subsection in the following only *one* feature is considered. Hence, transiently, the index i is omitted.

Moreover, at first a possible specification of *similarity* will be treated for *crisp* sets A and B.

Two *crisp* sets are equal, if they are *identical*, $A = A \cap B = B$ or, in equivalent form, if $A \subseteq B$ and $B \subseteq A$. A possible meaning of *similarity* can start with the conception that two sets are *similar*, if they are *approximately* equal. An interpretation of "approximately equal" could read that the subsets of the sets *outside* their intersection $A \cap B$ (the set of all elements belonging to both the sets) is "small" when compared with their union $A \cup B$ (the set of all elements belonging to at least one of the two sets). Now, the corresponding sets must be equipped with a measure of content: for a finite set this would be its *number* of elements, for a continuous set the *integral* over it. If the character of the set (discrete or continuous) should not be indicated, then the respective measure is called the *cardinality* card of the set. This way of speaking differs but essentially from the definition due to CANTOR, however, it is rather useful in the present context. In this manner a possibility to define the similarity $\text{sim}(A, B)$ of two crisp sets can be given by the following expression:

$$\text{sim}(A, B) = \frac{\text{card}\,(A \cap B)}{\text{card}\,(A \cup B)} \, . \tag{6.42}$$

This formula corresponds, by the way, with that coefficient suggested by SNEATH (1957) and mentioned already in Subsect. 6.1.2. Naturally, this is only an example for the variety of possibilities, with respect to further examples see e.g. BANDEMER/NÄTHER (1992).

This approach remains useful, if the sets A and B are allowed to be also *fuzzy* ones. In this case the common integral over the crisp set becomes an integral over the membership function, i.e. one takes as the cardinality the aera below this function, a rather comprehensible choice. With discrete sets one takes the corresponding sums over the membership values. One can apply here even other t-norms instead of the usual min-max variant for defining intersection and union. Moreover, instead of the just defined cardinality also other suitable measures can be used for the given purpose (see BANDEMER/NÄTHER (1992), where also some practical applications are shown).

If in the universe corresponding to a *single* feature a distance $d(u, v)$ is introduced, then this distance can be generalized via an extension principle

$\mu\,(x)$

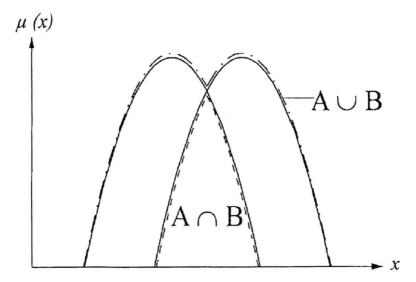

$A \cup B$

$A \cap B$

x

Fig. 6.3. Similarity of fuzzy sets, represented by the quotient of the areas below the membership functions of $\mathcal{A} \cap \mathcal{B}$ and $\mathcal{A} \cup \mathcal{B}$

(see Subsect. 3.2.4) to a distance $d(\mathcal{A}, \mathcal{B})$ between fuzzy sets. However, the transition from such a generalized distance to a similarity relation requires a repeated application of an extension principle and leads to fuzzy sets of type 2, as they are called, which will be considered below within another context.

Moreover, distances between fuzzy sets are also introduced basing on the difference of the membership functions (references see in BANDEMER/NÄTHER (1992)). Finally, there can be found also concepts and practical examples for a similarity notion with fuzzy functions in such a manner that the function value $f(z)$ at every argument point z is given as a fuzzy set $\mathcal{Y}(z)$ over a universe U_y.

If a fuzzy similarity relation over a universe U of a feature could be specified, then it is possible and sometimes more appropriate if the similarity of two fuzzy sets *in the sense of* $\mathcal{R}$ is expressed as a *fuzzy* set $\mathcal{S}$, which can be interpreted as a value of some linguistic variable SIMILARITY on the similarity scale $[0, 1]$. For the computation of $\mathcal{S}(\mathcal{A}, \mathcal{B}; \mathcal{R})$ an extension principle can be used, e.g.

$$\mu_S(z; \mathcal{A}, \mathcal{B}; \mathcal{R}) = \sup_{(u,v):\mu_R(u,v)=z} \min\{\mu_A(u), \mu_B(v)\} . \qquad (6.43)$$

This is a fuzzy set of type 2, this will be clear immediately from the interpretation of $S(\mathcal{A}, \mathcal{B}; \mathcal{R})$: The *fuzzy* similarity is expressed *fuzzily*, e.g. as a value of a linguistic variable. Hence $S(\mathcal{A}, \mathcal{B}; \mathcal{R})$ is called *fuzzily expressed fuzzy similarity* of $\mathcal{A}$ and $\mathcal{B}$ in the sense of $\mathcal{R}$. Although the procedure for computation of μ_S looks rather complicated, one can solve problems of classification by this means. Especially the procedure is practicable in the case, when the values of the linguistic variable (e.g. for a feature of objects) can be specified explicitly, e.g. by fuzzy sets $\mathcal{F}_1, \mathcal{F}_2, \ldots, \mathcal{F}_\nu$ over U *without* specifying $\mathcal{R}$ over this universe before. Then the fuzzy similarity values $(\mathcal{T}_1, \ldots, \mathcal{T}_\kappa)$ (over $[0,1]$) must be specified, by experts, only for all couples (F_r, F_s) of feature shapes, e.g.

$$S(\mathcal{F}_r, \mathcal{F}_s) = \mathcal{T}_{rs} \quad \text{with} \quad r, s \in \{1, \ldots, \kappa\} . \tag{6.44}$$

This (generalized to fuzzy sets) *fuzzy similarity relation* S can be used to construct, for every fuzzy set $\mathcal{A}$, *neighbourhoods* within the set of *all* fuzzy sets over U. So, a *crisp* set of fuzzy sets $J(\mathcal{A}; c_0, \mu_0)$ with

$$J(\mathcal{A}; c_0, \mu_0) = \left\{ \mathcal{B} : \sup_{z \geq c_0} \mu_S(z; \mathcal{A}, \mathcal{B}; \mathcal{R}) \geq \mu_0 \right\} \tag{6.45}$$

will be called a (c_0, μ_0)-*neighbourhood of* $\mathcal{A}$ where $c_0, \mu_0 \in (0,1]$ are the parameters of this neighbourhood. This neighbourhood contains all those fuzzy sets $\mathcal{B}$, which are (in the sense of $\mathcal{R}$) similar to $\mathcal{A}$ at least with the degree c_0 with a membership of at least μ_0. If one cannot or will not fix some μ_0 because this would not make sense, one can specify also a *fuzzy neighbourhood* of $\mathcal{A}$, e.g. by

$$\mathcal{J}_{c_0}(\mathcal{A}, \mathcal{R}) : \mu_J(\mathcal{B}, \mathcal{A}; c_0) = \sup_{z \geq c_0} \mu_S(z; \mathcal{A}, \mathcal{B}; \mathcal{R}) . \tag{6.46}$$

The membership value for $\mathcal{B}$ in the neighbourhood of $\mathcal{A}$ is additionally marked in Fig. 6.4. Over the set of all fuzzy sets over U the expression $\mathcal{J}_{c_0}$ is a fuzzy similarity relation.

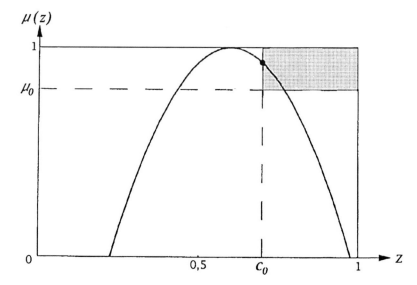

Fig. 6.4. Representation of a (c_0, μ_0)-neighbourhood of the fuzzy set $\mathcal{A}$

The application is especially practicable in the case mentioned above, if all occurring fuzzy sets are values of linguistic variables. In this case one has only to list all the couples of feature shapes, which belong each to the same fuzzy similarity value, e.g.

$$
\begin{aligned}
\mathcal{T}_1 \quad &\cdots \quad \{(r_{11}, s_{11}), \dots, (r_{1n_1}, s_{1n_1})\} \\
\mathcal{T}_2 \quad &\cdots \quad \{(r_{21}, s_{21}), \dots, (r_{2n_2}, s_{2n_2})\} \\
&\cdots \\
\mathcal{T}_\kappa \quad &\cdots \quad \{(r_{\kappa 1}, s_{\kappa 1}), \dots, (r_{\kappa n_\kappa}, s_{\kappa n_\kappa})\} .
\end{aligned}
\tag{6.47}
$$

Then only those membership curves of the $\mathcal{T}_t$ are to be confronted with the rectangle for the (c_0, μ_0)-neighbourhood, in the index set of which the index for the set $\mathcal{A}$ is contained. These will be, as a rule, rather few.

Finally, by a meaningful and systematic enlargement of this rectangle one has the possibility of a "clustering" procedure for the given sets around $\mathcal{A} = \mathcal{F}_0$.

In many cases, however, the handling of fuzzily expressed fuzzy similarities will be too confused and involved. Hence, it seems to be desirable to assign to the couples of fuzzy sets over U each a *scalar* value for solving classification problems, as it has been tried by ad-hoc proposals in the beginning of this section. These *similarity degrees* then define a fuzzy similarity relation over the set of all fuzzy sets over U. One of these proposals for a systematic construction of such similarity degrees with a given similarity relation $\mathcal{R}$ over U starts from multi-valued logic due to KLAUA (1966), (1966a) (see BANDEMER/NÄTHER (1992) and BANDEMER/GOTTWALD (1995)). E.g. one obtains by connecting with the algebraic product (see (3.31)) and using an "optimistic" interpretation of the similarity relation the similarity degree

$$
r_{\text{optalg}}(\mathcal{A}, \mathcal{B}; \mathcal{R}) = \sup_{(u,v)} \left\{ \mu_A(u) \cdot \mu_B(v) \cdot \mu_R(u, v) \right\} .
\tag{6.48}
$$

In analogy to the approach with $\mathcal{S}$ according to (6.43) also similarity degrees r of arbitrary origin can be used for establishing neighbourhoods of a fuzzy set $\mathcal{A}$. For every *crisp* similarity degree r_0 a *crisp* subset $J_r(\mathcal{A}; r_0)$ of the set of all fuzzy sets over U can be specified containing all those fuzzy sets, which are similar to $\mathcal{A}$ at least with a degree r_0:

$$
J_r(\mathcal{A}; r_0) = \{\mathcal{B} : r(\mathcal{A}, \mathcal{B}; \mathcal{R}) \geq r_0\} ,
\tag{6.49}
$$

where $r_0 \in (0, 1]$ is the parameter of the considered neighbourhood.

Naturally, also here a *fuzzy* neighbourhood $J_{r_0}(\mathcal{A})$ can be introduced. For this purpose the value $r(\mathcal{A}, \mathcal{B}; \mathcal{R})$ is to be interpreted as the membership value of $\mathcal{B}$ to the fuzzy neighbourhood of $\mathcal{A}$ (and vice versa).

6.3.2 The Use of The Concept for Classification

After this exposition on the specification of similarity for fuzzy sets now the data matrix (6.41)

$$\mathbf{X} = ((\mathcal{X}_{ij})) \tag{6.50}$$

is considered again, which is frequently called a *knowledge base*, because it contains also additional background knowledge by the fuzzy specification of the data.

From this $\mathbf{X}$ one obtains a similarity matrix for each feature F_i

$$\mathbf{S}_{jk} = ((\mathcal{S}(o_j, o_k))) \, , \tag{6.51}$$

which represents the fuzzily expressed similarity of the two objects with respect to the feature considered.

The neighbourhood of an object o_j is no longer required with respect to all possible fuzzy sets over U_i, but only with respect to the other objects from O or to a new object o_{N+1}, e.g.

$$J(\mathcal{X}_{ij}; c_0, \mu_0) = \{\mathcal{X}_{ik} : \sup_{z \geq c_0} \mu_S(z; \mathcal{X}_{ij}, \mathcal{X}_{ik}; \mathcal{R}) \geq \mu_0)\} \, , \tag{6.52}$$

as it was shown in Fig. 6.4.

If for *all* features $F_1, F_2, \ldots, F_t$ fuzzily expressed fuzzy similarities are specified, then the hypermatrix

$$\mathbf{S} = ((\mathcal{S}_{ijk})) \tag{6.53}$$

represents the *similarity structure* of the feature system of the knowledge base for all objects. This hypermatrix $\mathbf{S}$ can be used to tackle diverse problems of classification.

For fixed $i = i_0$ the matrix of fuzzy sets

$$\mathbf{S}_{i_0} = ((\mathcal{S}_{i_0jk})) \tag{6.54}$$

reflects the *fuzzily expressed similarity of all objects* with respect to the i_0-th feature. This matrix can be used to introduce *neighbourhoods of features* within the set of all features (see BANDEMER/NÄTHER (1992)).

For fixed object indices $j = j_0$ and $k = k_0$ the vector of fuzzy sets

$$\mathbf{S}(j_0, k_0) = (\mathcal{S}_{1j_0k_0}, \ldots, \mathcal{S}_{tj_0k_0}) \tag{6.55}$$

represents the fuzzily expressed similarity of the two objects o_{j_0} and o_{k_0} with respect to all features. Here *neighbourhoods of objects* can be introduced, e.g.

$$J(\mathbf{X}_j; c_0, \mu_0) = \left\{ \mathbf{X}_k : \quad \text{for all } i : \max_{z \geq c_0} \mu_S(z; \mathcal{X}_{ij}, \mathcal{X}_{ik}; \mathcal{R}) \geq \mu_0 \right\} . \tag{6.56}$$

In analogy to the proceeding with the fuzzy sets $\mathcal{S}_{ijk}$ also similarity degrees

$$r(\mathcal{X}_{ij}, \mathcal{X}_{ik}, R) = \mu_i(j, k) \tag{6.57}$$

can be used for the formation of neighbourhoods. For each i and for every similarity degree r_0 a crisp set of objects o_k can be determined containing all objects, which are similar to o_j with respect to the i-th feature at least to the degree r_0:

$$J_i(\mathcal{X}_{ij}; r_0) = \{\mathcal{X}_{ik} : \mu_i(j, k) \geq r_0\} . \tag{6.58}$$

By combining the specifications with respect to *all* features one obtains the hypermatrix

$$\mathbf{M} = ((\mu_i(j, k))) , \tag{6.59}$$

which represents the *similarity structure of the knowledge base* with respect to all objects and all features.

For *fixed* i_0 the matrix

$$\mathbf{M}(i_0) = ((\mu_{i_0}(j, k))) \tag{6.60}$$

reflects the *fuzzy similarity of all objects* with respect to the i_0-th feature. A *neighbourhood* of the i_0-th feature can now be determined e.g. by

$$J_F(i_0; \epsilon) = \{i : \text{ for all } j, k : |\mu_{i_0}(j, k) - \mu_i(j, k)| \leq \epsilon\} , \tag{6.61}$$

where $\epsilon > 0$ is the parameter characterizing the neighbourhood. In (6.61) implicitly the distance measure for matrices

$$d(\mathbf{E}, \mathbf{F}) = \max_{jk} |e_{jk} - f_{jk}| \tag{6.62}$$

is used. Naturally, also other distances are possible, if they reflect the opinion of the user on the *variedness* of the features within the context of the knowledge base. A smaller distance means then that the two features considered behave rather similarly with regard to the given set of objects and hence supply similar information on the variedness of the objects. This can be used for decisions, whether and which of the features can be omitted in future for being redundant, and, on the other side, it can form the starting point for an "interpolating" of missing values of features if necessary.

Moreover, the matrix $\mathbf{M}(i_0)$ can be used to evaluate the *discriminablility* of the i_0-th feature with respect to the given set of objects. If the elements μ_{i_0} for $j \neq k$ are all approximately equal to each other, then there is only little hope for finding any substructure of the objects by means of this feature. If, particularly, $\mathbf{M}(i_0)$ is the unit matrix, then the feature discriminates among the objects to the utmost, however, it does not demonstrate any non-trivial substructure, either.

Another possibility to evaluate the discriminability of a feature seems to be given by the SHANNON *entropy*

$$H(\mathbf{M}(i_0)) = -c \sum_{j,k} \mu_{i_0}(j, k) \ln(\mu_{i_0}(j, k)) , \tag{6.63}$$

where c is a normalizing constant depending on the elements of the matrix (see also BANDEMER/NÄTHER (1992)). If this entropy is large, then the loss in information is small, if that feature is omitted from the consideration. (Entropy is here a pure information theoretical value without any physical interpretation.)

For fixed $j = j_0$ and $k = k_0$ the vector

$$\mathbf{M}(j_0, k_0) = (\mu_1(j_0, k_0), \ldots, \mu_t(j_0, k_0)) \tag{6.64}$$

represents the *fuzzy similarity* of the two objects o_{j_0} and o_{k_0} and with respect to *all* features. A neighbourhood of the object o_{j_0} would be given by

$$J_r(j_0; r_0) = \{l : \min_i \mu_i(j_0, l) \geq r_0\}, \tag{6.65}$$

where r_0 is the parameter characterizing the neighbourhood. The so specified neighbourhood contains all objects of the knowledge base that are similar to o_{j_0} at least to the degree r_0 in *all* features. Obviously, the neighbourhoods can also be specified for only a part of the knowledge base and with different bounds for different features.

Finally, the similarity degrees $\mu_i(j, k)$ may be *aggregated* with respect to the features, e.g. by a suitable functional defining an *overall* degree of similarity between each two objects with respect to all features. This would lead back, in a certain manner, again to the introduction of an overall distance between the objects. Therefore the choice and interpretation of such an expression is controversial in a similar manner and with the same arguments as with that distance.

The neighbourhoods can now be used to solve problems usually, i.e. when the feature values are crisp, tackled by cluster analysis. This will be very clear in the case frequently met in practice that *typical* objects can be selected or constructed. Then the objects allocated by the corresponding neighbourhoods represent the equivalent to those clusters determined to each of the initial prototypes.

The practical solution of classification tasks for fuzzy feature values require, obviously, another structuring of computer software, however, this should not be a serious problem with the present state-of-the-art in computer science. The obvious advantage of such a proceeding is its narrow guide along the practical environment, such that at every time the user's knowledge of facts can be introduced into the consideration, being even demanded for application, whereas the "classical" solving procedures introduce mathematically motivated additional conditions, and the applying scientist is involved again not until for the assessment of the numerically found results, e.g. in cluster analysis, for the assessment of their meaningfulness.

7

Evaluation of Functional Relationships

As already introduced in Subsect. 2.4.1, a *functional relationship* is explained as a relationship between one (or several) dependent variable and one (or several) explanatory variable, which can be represented by a functional expression containing yet unknown parameters, which can assume values from a given index set. The determination of these free parameters should be executed by means of results from measurements or observations, actually, as a rule, only *approximately*. This proceeding is called *evaluation of a functional relationship* using experimental results.

The finding and fixing of such a special functional relationship (called the *setup*) is the decisive main problem in every kind of approximation from experimental results. Here the special knowledge of the applying scientist with respect to the environment of his investigation is required. One should read again in Subsect. 2.4.1 with regard to the problem of choosing a setup.

Compared with Chap. 2, however, a new aspect is introduced by founded assumptions on the genesis of the data used and their mathematical form and structure. These assumptions *increase* the *information content* of the data *essentially* and hence the *meaningfulness* of the statements on the connection (its sensibleness, precision, and reliability). Therefore, one should consider these assumptions thoroughly and check them carefully. Because all obtained statements are correct only on the *condition of the validity of all these assumptions*, their uncritical use can lead to euphoric assessments of the statements, to rough misinterpretations and even to uselessness of the obtained results. All the remarks in Chap. 1 with respect to data quality are also for the present chapter to a high degree and also with respect to the assumptions on the genesis of the data and their structure.

In the first section of this chapter the data are considered as realizations of random variables, in the second section at first the dependent variable is allowed to be fuzzy, and finally *all* the variables involved can assume fuzzy values.

7.1 Statistical Regression Analysis

7.1.1 Model Assumptions with Random Dependent Variables

As in Chap. 2 n given values y_i of the dependent variable at the corresponding points $\mathbf{x}_i$ with

$$\mathbf{x}_i = (x_{1i}, \ldots, x_{ki}) \tag{7.1}$$

form the starting point of the information situation. In contrast to Chap. 2 the arguments here are always considered multidimensional, because this case is the most frequent one in practical application. For the sake of simplification of the presentation the dependent variable is taken *one-dimensional*, the generalization to the multidimensional case is possible, on principle, but rather more complicated in its treatment. With respect to this *multivariate* case it is referred to textbooks (see e.g. MARDIA/KENT/BIBBY (1979)).

The problem will consist in the specification of a function

$$y = g(\mathbf{x}) , \tag{7.2}$$

which connects the dependent variable y with the arguments $x_j; j = 1, \ldots, k$, the components of the vector $\mathbf{x}$, now called *explanatory variables*.

Up to now the points $(\mathbf{x}_i, y_i)$ were considered as given *exactly*. Now some *additional assumptions* are introduced.

The *first assumption* enlarges the *information situation*. The values y_i of the dependent variable are allowed to be influenced by *observational errors*. However, the values of the explanatory variables x_{ji} should be given still *exactly*, or, formulated more realistically: the observational or adjusting errors of the explanatory variables x_j should be allowed to be *neglected* when compared with the observational and measurement errors of the dependent variable y. The typical case in application for this assumption is the technological measurement of a technologically defined quantity in dependence on very precisely adjustable influencing quantities.

The *second assumption* defines the kind of data genesis. The values of the dependent variable y_i are taken as *realisations of random variables* Y_i, which can be splitted each into two terms

$$Y_i = Y(\mathbf{x}_i) = g(\mathbf{x}_i) + \epsilon(\mathbf{x}_i) , \tag{7.3}$$

the value $g(\mathbf{x}_i)$ of the required function at the point $\mathbf{x}_i$ and the realization $\epsilon(\mathbf{x}_i)$ of a *random* error.

The function $g(\mathbf{x})$ is called also *response surface*. Hence the random variable Y is then the *response variable* and its realizations are the *response values*.

The *third assumption* demands that the random error ϵ does not influence the values of the response variable *systematically*:

$$E\epsilon\,(\mathbf{x}) = 0 \tag{7.4}$$

for all possible $\mathbf{x} \in B \subseteq I\!R^k$.

In order to make the mathematical treatment easier a *fourth assumption* is introduced: The random errors of the different measurements should *not* influence each other. This is really a demand for statistical independence, but it is moderated, as a rule, to a demand for *uncorrelatedness*

$$\text{cov}(\epsilon(\mathbf{x}_i), \epsilon(\mathbf{x}_l)) = \text{E}[\epsilon(\mathbf{x}_i), \epsilon(\mathbf{x}_l)] = 0 \quad \text{for} \quad i \neq l \tag{7.5}$$

which is, as is well known, equivalent in the case of normally distributed errors.

Moreover, frequently a *fifth assumption* is put that the precision of the measurements should be independent of the point, where the measurement takes place, i.e.

$$\text{D}^2\epsilon(\mathbf{x}_i) = \sigma^2 \quad \text{for all} \quad i = 1, \ldots, n. \tag{7.6}$$

The fourth and fifth assumption can be moderated, though then other assumptions are necessary on the kind of the stochastic dependence or of the kind of the dependence of the variance on the points of measurement, respectively. The methods for such cases should check these assumptions, because they influence the results *explicitly*.

As in the case of approximation in Chap. 2, without any assumption on the function $g(\mathbf{x})$ no conclusions can be drawn with regard to the values of the function *between* the points of measurement. One needs, as in that case, a functional relationship

$$y = \eta(\mathbf{x}; \mathbf{a}) \tag{7.7}$$

with free parameters $\mathbf{a} = (a_1, \ldots, a_m)$, which is called usually a *setup* in this context. In view of the later treatment in numerical procedures now also the parameters are handled as a *vector*.

Because it leads to simple mathematical tasks for a solution, the *linear setup* is particularly popular. Linearity refers here only to the free parameters $a_s; s = 1, \ldots, m$. One starts in this approach with m *known* functions f_s, which are *linearly independent* over the considered domain B, and combine them linearly by the free parameters:

$$\eta(\mathbf{x}; \mathbf{a}) = \sum_{s=1}^{m} a_s f_s(\mathbf{x}). \tag{7.8}$$

This approach was already considered, for only *one* argument x, with approximation in Chap. 2. As explained there, the choice of a setup depends essentially on the *aim* of the investigation. For an investigation of scientific and technological connections also non-linear (in the parameters) setups are of high interest (e.g. for saturation processes), if they can be founded by facts. If but only a comfortable presentation is of interest, perhaps only in a small subdomain, then one chooses a setup, as a rule, which is linear in the parameters.

As a motivation then the possiblity of a TAYLOR- or FOURIER-representation, as with approximation, will be effective. Also the recommendation to use orthogonal functions may be helpful here.

For a rational interpretability of the results obtained by the methods from mathematical statistics (estimation, prediction, test decisions) a *sixth assumption* is necessary: It should exist a particular value $\mathbf{a}_0$ of the parameter vector with

$$g(\mathbf{x}) = \eta(\mathbf{x}; \mathbf{a}_0) \,, \tag{7.9}$$

i.e., the unknown response surface to be found should be a particular function of the functional relationship, or – with other words – the response surface must have exactly the form of the functions of the setup. In this case this setup is called a *true setup*.

The treatment of the approximation and the prediction problem for the function g on the assumptions mentioned in this section (and of generalizations and leading on investigations based on these assumptions) is called *regression analysis*. The naming took place long ago in connection with an anthropological problem, the presentation of which can be omitted here.

Problems in connection with the assumptions presented in this subsection are considered again in Subsect. 7.1.3, after a short presentation of some methods of mathematical statistics.

7.1.2 The Problem of Estimation

The six assumptions specified in the preceding subsection are used now to formulate the problem:

Specify the unknown function g approximately by the data $(\mathbf{x}_i, y_i)$ as a problem of *mathematical statistics* and estimate the unknown parameter vector $\mathbf{a}_0 = (a_{10}, \dots, a_{m0})^\tau$ by means of the given sample vector $(y_1, \dots, y_n)^\tau$.

The vectors are used here as column vectors, as usual in this context, in the following; hence a transposition sign is necessary, when they occur as row vectors within the text.

First the simple case of a *linear setup* is considered. In this case the formulation of the task and the results by means of matrix theory is advisable.

For the sake of simplicity, restricting the set of possible estimations to those of *linear forms* of the sample

$$\hat{a}_s = \sum_{i=1}^{n} c_{is} y_i + c_{0s} \tag{7.10}$$

and putting the additional condition of *unbiasness* (see Subsect. 4.2.2 and (7.18) below)

$$\mathbf{E}\hat{\mathbf{A}} = \mathbf{a}_0 \,, \tag{7.11}$$

then the "best" estimation is obtained by the method of least squares (more precisely: the method of the least sum of squares). This method was introduced already in Sect. 2.3, though for only *one* explanatory variable x. In contrast here a *vector* $\mathbf{x}$ of explanatory variables is allowed and for the response variable y a stochastic background is considered.

With the given realizations y_i and the true linear setup (7.8) and (7.9) for the unknown function g the optimization problem of the method of least squares has the mathematical form

$$Q(\mathbf{a}) = \sum_{i=1}^{n} \left(y_i - \sum_{s=1}^{m} a_s f_s(\mathbf{x}_i) \right)^2 = \min_{\mathbf{a}} . \tag{7.12}$$

In usual manner by partial derivation of $Q(\mathbf{a})$ with respect to the components of $\mathbf{a}$ and nullifying of the derivatives the following system of linear equations is obtained

$$\sum_{s=1}^{m} a_s \sum_{i=1}^{n} f_s(\mathbf{x}_i) f_1(\mathbf{x}_i) = \sum_{i=1}^{n} y_i f_1(\mathbf{x}_i)$$

$$\cdots \tag{7.13}$$

$$\sum_{s=1}^{m} a_s \sum_{i=1}^{n} f_s(\mathbf{x}_i) f_m(\mathbf{x}_i) = \sum_{i=1}^{n} y_i f_m(\mathbf{x}_i) .$$

With the abbreviations

$$\begin{aligned}
\mathbf{F} &= ((f_s(\mathbf{x}_i))); \quad s = 1, \ldots, m; \quad i = 1, \ldots, n; \\
\mathbf{y} &= (y_1, \ldots, y_n)^\tau \\
\mathbf{a} &= (a_1, \ldots, a_m)^\tau
\end{aligned} \tag{7.14}$$

it can be written clearly as

$$\mathbf{F}^\tau \mathbf{F} \mathbf{a} = \mathbf{F}^\tau \mathbf{y} . \tag{7.15}$$

If the system matrix $\mathbf{F}^\tau \mathbf{F}$ is *non-singular*, i.e. it possesses an inverse matrix, then the system of equations (7.15) can be resolved for $\mathbf{a}$ formally and yields the solution $\hat{\mathbf{a}}$, the wanted estimating value:

$$\hat{\mathbf{a}} = (\mathbf{F}^\tau \mathbf{F})^{-1} \mathbf{F}^\tau \mathbf{y} . \tag{7.16}$$

Necessary conditions, on which the matrix is *non-singular*, are the linear independence of the set functions over the considered domain and that the observations took place at points $\mathbf{x}_i$, of which at least m were *different* from each other.

Generally, the vector of the response values is a random vector $\mathbf{Y}$ (the procedure should be valid for all possible realizations), hence the estimator is a random vector

$$\hat{\mathbf{A}} = (\mathbf{F}^\tau\mathbf{F})^{-1}\mathbf{F}^\tau\mathbf{Y} . \tag{7.17}$$

Numerically, the vector $\hat{\mathbf{a}}$ is provided by a computer. As mentioned already in Chap. 2 also here a *numerical* error is supervened by the computation, which has, however, *nothing in common* with the random error, which occurred in specifying the values y_i. The scale of the numerical error depends on the equation system, more precisely, on the *condition* of the matrix $\mathbf{F}^\tau\mathbf{F}$. The effect was demonstrated already in Subsect. 2.3.1 by the example of a glancing intersection of two straight lines. Also in the present case the advice remains valid to assess this numerical error in every case. This can be effected by comparing the solutions of the system of equations, when the "inputs" are changed several times numerically to only an insignificant extent each time.

A purely statistical problem, however, is the assessment of *precision of the estimation*, i.e. the question, how the measurement errors ϵ propagate themselves to the estimation. Let be $\delta = (\delta_1, \ldots, \delta_m)^\tau$ the random vector of the deviations from the "true value" $\mathbf{a}_0$:

$$\mathbf{a}_0 + \delta = \hat{\mathbf{A}} \tag{7.18}$$
$$= (\mathbf{F}^\tau\mathbf{F})^{-1}\mathbf{F}^\tau\mathbf{Y}$$
$$= (\mathbf{F}^\tau\mathbf{F})^{-1}\mathbf{F}^\tau(g(\mathbf{x}) + \epsilon)$$

with $g(\mathbf{x}) = (g(\mathbf{x}_1), \ldots, g(\mathbf{x}_n))^\tau$ and $\epsilon = (\epsilon(\mathbf{x}_1), \ldots, \epsilon(\mathbf{x}_n)^\tau)$. Then, on account of the unbiasedness (7.11) one obtains by the simple result

$$\delta = (\mathbf{F}^\tau\mathbf{F})^{-1}\mathbf{F}^\tau\epsilon \tag{7.19}$$

a connection between the estimation errors δ and the measurement or input errors ϵ. Even if the input errors are uncorrelated, the estimation errors, in general, will be *correlated*. Hence, for an assessment of the estimation precision one uses quantities, which are derived from the covariance matrix of the estimation error vector

$$\mathbf{B}_\delta = ((\mathrm{E}(\delta_s\delta_r))); \quad s, r = 1, \ldots, m . \tag{7.20}$$

In the main diagonal of this matrix one finds the variances $\mathrm{D}^2\delta_s$ of the estimators $\hat{\mathbf{A}}_s$ of the parameters a_s. If the measurement errors are uncorrelated with a constant variance σ^2, then one obtains immediately from (7.19)

$$\mathbf{B}_\delta = \sigma^2(\mathbf{F}^\tau\mathbf{F})^{-1} . \tag{7.21}$$

If, moreover, the matrix in brackets is a diagonal matrix, then also the components of the estimation are uncorrelated and one can read the precision of the parameter estimation directly. Therefore, if the points $\mathbf{x}_i$, where the measurement should take place, can be *chosen* beforehand, then one can try to choose them, e.g., in such a manner that one will obtain, with the given functions f_s, a diagonal matrix $\mathbf{B}_\delta$. This is a problem of *statistical experimental design* (see e.g. BANDEMER/NÄTHER (1980)).

The given estimator $\hat{\mathbf{A}}$, according to (7.17) and on the assumptions from Subsect. 7.1.1, is the best one in the following sense that every other *linear unbiased* estimation has a "larger" covariance matrix $\mathbf{B}_\delta$ (in the sense of the partial ordering of positive definite matrices, see some textbook on matrix theory).

7.1.3 Discussion of the Model Assumptions

The statements of Subsect. 7.1.2 are valid only on the assumptions of Subsect. 7.1.1. Now it will be investigated what will happen, if some of them are no longer valid, and how these assumptions can be checked or guaranteed.

With the *first* assumption is was presupposed that the points $\mathbf{x}_i$, where the measurements took place, are relatively precise when compared with the response values y_i. If $\mathbf{x}_i$ as well as y_i are afflicted by observation errors being not neglectable, then $(X_1, \ldots, X_k, Y)$ is a random vector. This forms the starting point for the general regression problem with random variables (called the regression model of second kind) and for the regression analysis with errors in variables. Subsect. 7.1.5 will be devoted to these problems.

By the *second* assumption it was demanded that the random error ϵ occurs as an additive term ("absolute" error). Frequently, however, the "relative" error will play a role, i.e. the connection is a multiplicative one:

$$Y(\mathbf{x}_i) = g(\mathbf{x}_i)\epsilon\,(\mathbf{x}_i)\,. \qquad (7.22)$$

Formally a transition is possible then, in turning to the logarithm on both sides and putting down the problem again to an additive one:

$$\log Y(\mathbf{x}_i) = \log g(\mathbf{x}_i) + \log \epsilon\,(\mathbf{x}_i) \qquad (7.23)$$

changing the notation appropriately. However, this near at hand proceeding can be treacherous. The transformed problem can show totally unexpected results. Apart from the fact that all occurring quantities must be positive, of which the logarithms should be taken, the *distances* on the y-axis are changed radically. It can be disastrous, first of all in the case, when the input errors are really essentially additive terms.

By the *third* assumption it was demanded that the measurement errors do not provoke *systematic* deviations in the results. In practice preferences can but happen at all. In measuring, e.g., of inner calibres of pipes the results are more likely too short than too long.

The cases of weak data quality, already discussed in Chap. 1, go also in this category of hurting assumptions. Hence the warnings and hints to be given here are essentially repetitions: A careful expertly done examination of the data material is always advisable.

Again it is warned of procedures, which realize and eliminate potential *outliers* (lying outside the bulk of the data) automatically. If procedures for realizing outliers are applied at all, so these must present the realized potential outliers to the handling scientist first. Starting from his expertly point of view and his technological possibilities the scientist will decide, what should be done: Whether the measurement or observation can or will be repeated, or the value can be corrected or left as a valid value, or, finally, it must be deleted after all. For interest of new insight it is advisable in any case to investigate the causes of the unexpected or seeming deviation with a potential outlier.

If one reckons already with a certain percentage of outliers, which will be "harmless" with respect to the statements of interest, then it might be useful to apply procedures, which are called "robust against outliers". Such procedures weigh the potential outliers according to the degree of their single "being outside" by factors smaller than 1 and hence decreasing their influence on the statements (e.g. estimations) respectively.

In general, the advice is repeated here that the scientist dealing with the data should require information as precise as possible on *how* the data are gathered, best by personal close look.

Finally, *residual analysis*, as presented shortly at the end of this subsection, can give some insight into possible hurtings of the assumptions.

With the *fourth* assumption the *independence* (or at least the *uncorrelatedness*) of the random measurements was required. If this assumption is hurted *systematically*, then one must know this *system*, when statements on the connection are to be made. The system can also be a stochastic one. The random response variable Y needs then a stochastic model, either a random process (e.g., when time is the only explanatory variable) or a random field (e.g., when coordinates of points are the explanatory variables). Models of this kind are beyond the framework of the present book, because they need an essentially more profound background. With respect to an important special case it is referred to Subsect. 7.1.6.

With regard to *time series* see, e.g., BOX/JENKINS (1970).

By the *fifth* assumption it was demanded that the variance is *constant* over the whole domain of the explanatory variables. But the variance can change, e.g., when the measurement scale measures the quantities only after a transformation, e.g., by their logarithms. When already results from earlier investigations are available or when by preliminary investigations the dependence of the measurement precision on the explanatory variables was made clear, e.g., by residual analysis, then the original formula of the method of least squares can be modified to obtain again good estimates. Let be $\sigma^2(\mathbf{x}_i)$ the variance at the respective point $\mathbf{x}_i$, then the modified optimization problem reads

$$\sum_{i=1}^{n} \sigma^{-2}(\mathbf{x}_i) \left(y_i - \sum_{s=1}^{m} a_s f_s(\mathbf{x}_i) \right)^2 = \min_{\mathbf{a}} . \tag{7.24}$$

In the software the variance values should be called up, as a rule, and taken respectively into account in the computation.

Finally, the *sixth* assumption is usually the most important one: The chosen setup should be true. If there is some factually supported uncertainty or doubt with respect to the setup, then this is the starting point for a *model discrimination*, as it is called usually. One is faced with this problem in different manner, whether the setup is expertly supported or a purely approximating setup.

With expertly supported setups one wishes to have a closed analytical form for the relationship, which should be used *always in future* as a scientifically based knowledge. It can happen, however, that for the given situation there exist *several* supported theories or assumptions on the (causal) connection, which each leads to an expertly supported setup. In an investigation it should be decided, which of these setups does justice to the given data material in the best way. This one obtains then, in the future, the greatest factual confidence.

A simple decision method for one of the setups is the following. For each of the setups the parameters are estimated according to the method of least squares and then that setup is chosen, for which the smallest residual sum $Q_{\min}$ is reached. But this method is problematic. It does *not* decide on the *truth* of the setup and of the theory basing it, as it is frequently desired. Moreover, setups with *more* parameters are preferred, because the approximation quality increases, as a rule, with the number of free parameters. The decision on an expertly supported setup should be, as in the last instance, always a problem of the applying scientist.

If the observational points can be chosen within a certian framework arbitrarily, method of statistical experimental design can be used to increase the power of discernment of the method of model discrimination.

With approximating setups the situation is quite different. Here a setup should be chosen, which, on the one side, represents the given data quite well, and on the other side, does not contain too many parameters. A later modification considering further data is not excluded. In this case the *model error*, e.g.

$$\min_{\mathbf{a}} \max_{\mathbf{x}} |g(\mathbf{x}) - \eta(\mathbf{x}, \mathbf{a})| \,, \tag{7.25}$$

should have the scale of the measuremet error σ. Instead of the maximum of the absolute value of the difference, as in (7.25) one could consider the integral over the square $(g(\mathbf{x}) - \eta(\mathbf{x}, \mathbf{a}))^2$ over the domain B and try to bring it into the scale of the variance σ^2. Especially interesting is a proceeding beginning with a simple setup (e.g., linear in the explanatory variables) and enlarging it succesively by further terms, until a satisfying approximation of the data is reached. This is the usual procedure in statistical designing in the sense of BOX (see with regard to further literature also BANDEMER/NÄTHER (1980)).

In every case it should be warned of an unsystematic trial and error approach, as it is offered by some software tools. From a catalogue of standard functions (polynomials, trigonometrical functions, logarithms, power functions and other exponential functions, up to "exotic expressions") terms are selected and compiled to setups, until the residual sum with the given data is sufficiently small. The result is almost never factually interpretable and always numerically unstable; slight changes in the data or additional measurements lead possibly to totally different setups.

Also the frequently for this purpose recommended *method of principal components* (see, e.g., LAWLEY/MAXWELL (1971) with respect to the problem and the procedure) is not without its problems as a method to choose a setup. On the one side the result depends considerably on the data, on the other side, the so created linear combination of partial setup functions, turning out as *new explanatory variables,* can frequently interpreted in the real application case only very difficultly. It needs really a well investigated background, when the method of principal components should yield a valuable result.

A tried method for a heuristical check of certain assumptions is supplied by *residual analysis,* although it represents, naturally, also no universal remedy and needs a supporting expertly interpretation and examination. From the "estimated errors"

$$\hat{\epsilon}_i = y_i - \eta(\mathbf{x}_i, \hat{\mathbf{a}}) \,, \tag{7.26}$$

which are connected by the estimating equations (hence they are no longer stochastically independent) possible hurtings of the assumptions are inferred. Usually for this purpose extensive software packages of statistical data analysis are used, which can visualize also multidimensional data via projections.

In the following two simple cases are demonstrated.

In Fig. 7.1 the estimated errors show the *same* sign over a whole subdomain of the x-axis. In this subdomain something is wrong with the chosen setup; possibly, in the present case, a quadratic term is missing in the setup.

In Fig. 7.2 the scale of the estimated errors changes with increasing x. Here the assumption demanding equal variance is hurted. Squaring the values

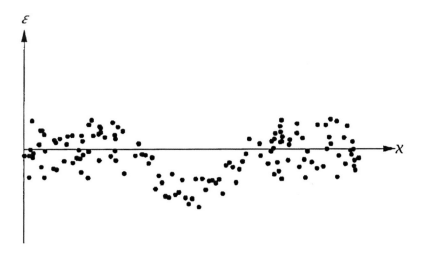

Fig. 7.1. Estimated error values from a sample drawn over the domain of the explanatory variable

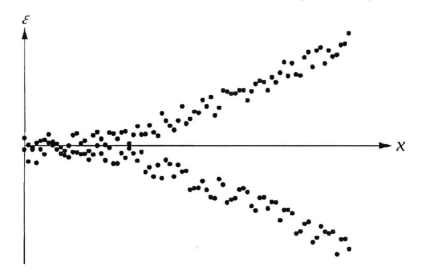

Fig. 7.2. Estimated error values from a sample drawn over the domain of the explanatory variable, another example

$\hat{\epsilon}_i$ and fitting a function over the domain yields an impression of the function $\sigma^2(x)$ to be used in the modified method of least squares.

Also drawing the estimated errors over y yields sometimes hints on a possible multiplicativity of the measurement errors.

The common software for data analysis offers further procedures of residual analysis also for the multidimensional case.

7.1.4 Further A-Priori Knowledge and Assumptions

According to the "golden rule" of statistics (What of a-priori knowledge does not exists or is not used, must be compensated by a higher expenditure of measurement and observation) further a-priori knowledge on the response surface can be used, either to increase precision and certainty of estimation or to decrease expenditure of observation.

If, e.g., certain points are known, where the response surface *must* assume known values, for factual reasons, then the setup can be reduced or the pointwise connections can be introduced as restrictions into the optimization problem. As an example: if $g(0) = 0$, then the constant in the setup can be omitted. Moreover, also given bounds, supported by facts, can be taken into account, e.g., $g(\mathbf{x}) \geq 0$ oder $g_u(\mathbf{x}) \leq g(\mathbf{x}) \leq g_o(\mathbf{x})$. Hints to procedures for these cases are contained, e.g., in TOUTENBURG (1982).

For specifying of a *confidence estimation* with

$$P_{\mathbf{a}_0}\Big(C(Y_1, \ldots, Y_n) \ni \mathbf{a}_0\Big) \geq 1 - \alpha \,, \tag{7.27}$$

i.e. a random domain $C(Y_1, \ldots, Y_n)$, which covers the true parameter value $\mathbf{a}_0$ with a probability of at least $1 - \alpha$, an additional assumption on the distribution of the observational errors ϵ is necessary.

Usually as a *seventh* assumption it is required that ϵ is *normally distributed* or at least *approximately* normally distributed. On this condition confidence regions for the unknown parameter and the unknown response surface make sense. If the assumption is hurted, then the regions are, as a rule, at least much too optimistic, sometimes, however, even totally senseless. An examination, whether the assumption is valid can be effected (conditionally) by residual analysis. It would be ideal, but too expensive, to have preliminary investiagations with several repeated measurements at each of the points. For *large* n the formulae for confidence regions are relatively robust against hurtings of the seventh assumption.

If observations or measurements are to be carried out very often, e.g. as an accompanying routine, or if the connection is already investigated rather well, one succeeds sometimes in specifying the theoretical or empirical a-priori knowledge by a frequency or probability distribution for the coefficients. This is a meaningful use of today nearly everywhere available "data heaps". BAYESIAN inference as already presented in Sect. 4.3 can be applied also in the present case of evaluation of functional relationships.

First an a-priori distribution for the vector $\mathbf{a}$ is to be specified, either as a histogram or by exploitation of experts' opinions. Practically this is mostly carried out gradually. In the first step an expectation vector $\mathbf{a}_B$ is fixed, either by averaging or by taking an assumed centre, around which the values scatter. Then a covariance matrix $\mathbf{B}_{\mathbf{a}_B}$ is fixed, either estimated from available data material or specified by factual considerations. Finally, a type of distribution is to be chosen yet. The BAYES' theorem yields then an a-posteriori distribution for the parameters and hence also for the response surface.

If the setup is true and *linear*, the problem proves to be essentially simpler. Then already expectation vector $\mathbf{a}_B$ and covariance matrix $\mathbf{B}_{\mathbf{a}_B}$ suffice to obtain a linear a-posteriori estimation for the true parameter vector $\mathbf{a}_0$. This linear BAYES' estimator, as it is called, has the form

$$\hat{\mathbf{A}}_B(\mathbf{Y}) = [\mathbf{F}^\tau \mathbf{F} + \mathbf{B}_{\mathbf{a}_B}^{-1}]^{-1} [\mathbf{F}^\tau \mathbf{Y} + \mathbf{B}_{\mathbf{a}_B}^{-1} \mathbf{a}_B] . \tag{7.28}$$

Comparing this estimator with the common estimator according to the method of least squares, one realizes that the corresponding expressions of the a-priori distribution are each *added* to those of that estimator. This makes it possible to come to reasonable estimates of the parameters even if the normal system matrix $\mathbf{F}^\tau \mathbf{F}$ is singular. Hence for a BAYES' estimator one gets by, as a rule, on a smaller observation expenditure, this is a result of the introduction of a-priori knowledge, i.e. additional information.

With the same observation expenditure the BAYES' estimator improves the quality of the estimation statement. This estimator is also relatively robust against smaller misspecifications of the a-priori quantities. But it is, naturally, a *preconceived opinion* about the parameters, this can lead to misinterpretations of the observation results. Hence it is suggested to keep always in mind the consistency of the a-priori assumptions with the observation results. There exist also tests, by which the consistency can be "checked" , however, it is warned to appley them uncritically. If it is possible, the dependence of the estimator on the a-priori distribution should remain transparent (see e.g. formula (7.28)). Further details of this field of problems see, e.g., PILZ (1991).

Sometimes it is possible to interprete also bounds for the response surface in the sense of BAYESIAN theory (see e.g. BANDEMER/PILZ/FELLENBERG (1986)).

7.1.5 Random Influence in All Variables

Up to now it was assumed that only the response variable y is a random variable, i.e. the uncertainty in $\mathbf{x}$ may be neglected. Now the case is considered that this assumption is no longer valid. With respect to this problem there are at least three different approaches.

In the *first* approach *all variables* involved, response variables as well as explanatory variables, are considered as *random variables*. For a presentation of this approach the simplest case of one random reponse variable Y and one random explanatory variable X will suffice.

So, starting point is the random vector (X, Y). The curve of the expected values of Y for each fixed x

$$E(Y|X = x) = g_Y(x) \tag{7.29}$$

and correspondingly

$$E(X|Y = y) = g_X(y) \tag{7.30}$$

are called the *regression lines* of the common distribution of X and Y. In general, they are different from each other and allow statements only in *one* direction each, either on the behaviour of Y with given x, or on that of X with given y. Conclusions in the each opposite direction, as with usual functions e.g. by an inverse function, are impossible.

Without any knowledge about the common distribution these notions are, as a rule, of only little use. Naturally, one could consider the common frequency distribution. However, for well-founded conclusions the number n of available realizations should be rather large.

There exists an "approximation approach", in which both the regression lines are approximated each by a linear setup, the free parameters of which are to be estimated then. The numerical procedure is identical with that for *non-stochastical* arguments x (with $g_Y(x)$) and y in the other case. In interpretation of the results of regression analysis, however, the model background is to be taken into account. This kind of regression with random variables is also called *regression of second kind*.

A *second* approach is the regression for functional relationships with *errors in variables*. The assumptions on the relationship and on the nature of the variables involved are similar to those put for regression with fixed and known values of the explanatory variables:

The unknown function $g(\mathbf{x})$, which describes the relation, is assumed continuous and even developable into a TAYLOR-series. A *true* setup $\eta(\mathbf{x}, \mathbf{a})$ let be known. The observatorial errors be all *additive*:

$$Y_i = y_i + \epsilon_i \qquad (7.31)$$
$$X_{ji} = x_{ji} + u_{ji} . \qquad (7.32)$$

The functional relationship be valid *exactly* for the *true* values, i.e.

$$y_i = g(\mathbf{x}_i) . \qquad (7.33)$$

The deviations ϵ_i and u_{ji} let be realizations of continuous random variables with expectation value zero each.

The given methods differ from each other by *further* assumptions.

If the deviations are *stochastically independent of each other* (further assumption a)) and if the variances of these deviations σ_Y^2 and σ_j^2 for Y and the corresponding component X_j of $\mathbf{X}$, respectively, are *known* (further assumption b)), then the free parameters $\mathbf{a}$ of the setup and the true observation points $\mathbf{x}_i$ can be estimated by a modified method of least squares:

$$Q_F(\mathbf{a}, \mathbf{x}_1, \ldots, \mathbf{x}_n)$$
$$= \sigma_Y^{-2} \sum_{i=1}^{n} \left(Y_i - \eta(\mathbf{X}_i; \mathbf{a}) \right)^2 + \sum_{j=1}^{k} \left[\sigma_j^{-2} \sum_{i=1}^{n} (X_{ji} - x_{ji})^2 \right] = \min . \quad (7.34)$$

There are methods to eliminate the x_{ji} in order to reduce the problem, because the dimension of the problem increases with the number of observation points.

Another approach suggests to divide the data material into groups and to estimate the parameter vector $\mathbf{a}$ from the centres of gravity of the single groups.

A survey of the problem with references can be found, e.g., in SCHMERLING/BANDEMER (1985).

Finally, methods of local approximation can be adapted also to this case (see Sect. 2.3).

The next subsection is devoted to an interesting and for application important special case of "local" estimation, if the observations are *stochastically dependent* on each other.

7.1.6 Local Regression in a Random Field

A special model, which plays its role in deposit geometry and which led to the development of a branch of *geostatistics* (see, e.g., CRESSIE (1991)), considers a *field function*, e.g., in the three-dimensional natural space

$$v = g(x, y, z) . \tag{7.35}$$

In this context x may indicate the length, y the width, and z the depth as corresponding coordinates, whereas v can indicate, e.g. the contents of ash, metal, or water. Actually, such a function can assume only the values 1 or 0, when considered on the level of single points: either the point lies within a corresponding particle or not. However, the value v at a single certain place (x_0, y_0, z_0) can be "measured" only as the *mean* taken from a "specimen" $P(x_0, y_0, z_0)$ of *finite* extent with fixed shape, quantity, and orientation in space around the point:

$$g_P(x_0, y_0, z_0) = \int_{P(x_0, y_0, z_0)} g(x, y, z) \mathrm{d}x\, \mathrm{d}y\, \mathrm{d}z , \tag{7.36}$$

perhaps divided by the volume of the specimen

$$\int_{P(x_0, y_0, z_0)} \mathrm{d}x\, \mathrm{d}y\, \mathrm{d}z . \tag{7.37}$$

It is also possible to introduce a weight function $w(x, y, z)$ within the single corresponding integrals.

Such a function $g_P(x, y, z)$ is called usually a *regionalized variable* in considering the variation of the specimen P with respect to shape, quantity, and orientation (see MATHERON (1965)).

The mathematical task consists in the local approximation of the function g_P, in order, e.g., to estimate the quantity of an interesting substance within a given solid. Let be B_0 a solid planned for being mined (a "block"). Because the specimens P are small when compared with the extent of the block, the quantity of the interesting substance within the block, $J(B_0)$, is obviously

$$J(B_0) = \int_{B_0} g(x, y, z) \mathrm{d}x\, \mathrm{d}y\, \mathrm{d}z = \int_{B_0} g_P(x, y, z) \mathrm{d}x \mathrm{d}y \mathrm{d}z , \tag{7.38}$$

if g_P was normalized by the integral (7.37). Another problem is the interpolation or approximation of g_P within a region V.

In both cases one wishes to "composite" the expression by *linear functions* of certain "measurement values" $g_P(x_i, y_i, z_i)$, e.g. by

$$\hat{J}(B_0) = \sum_i c_i g_P(x_i, y_i, z_i) \tag{7.39}$$

$$\hat{g}_P(x, y, z) = \sum_i d_i(x, y, z) g_P(x_i, y_i, z_i) . \tag{7.40}$$

The *most important* model idea for a statistical handling of these tasks consists in the following *assumption*:

The function g_P is, over its domain $B \subset \mathbb{R}^3$, a realization of a *random field* with

$$G_P(x, y, z) = g_0(x, y, z) + \epsilon(x, y, z) , \qquad (7.41)$$

where g_0 is called the *trend part* and ϵ the *random part*. Note the analogy with the regression model.

The application of this model is not unproblematic. Therefore some of these problems will be discussed here, because they find their expression in the course of computation explicitly.

Firstly, the splitting of the regionalized variable into a trend part and a random part depends essentially on the chosen *scale* of the investigation. In the two-dimensional case, local differences in the elevation of the ground are assessed by a landscape gardener in another way than by a road construction planner or even by a carthographer. What is a sytematic trend to be taken into account yet for the one of them can be already a random deviation for the other one. So, the *choice of the scale* depends on the intended aim and the context, removes arbitrariness and specifies what should be considered as randomness. In this way the statement is confirmed here, which was already held in Subsect. 4.1.1 that the applying scientist decides what should be considered as a matter of chance in his problem.

If one has decided, in this way, for a stochastic model idea, immediately a second problem arises: The given deposit is *only one* realization of the hypothetically assumed random field. Probabilistic statements are conclusions on *varieties* of possible realizations. Statistical conclusions from *only one* realization are, as a rule, highly questionable. There are different interpretations tried to escape from this dilemma, so, e.g., the assumption that the different blocks, in which the deposit is partitioned are realizations of some "random block". This supports the useful recommendation that the model should be considered as *valid* only *locally*.

From the definition of a regionalized variable above one realizes clearly that "neighbouring values" can be *not* independent of each other, hence that also random deviations ϵ will show a dependence. One expects an improvement of the quality of statistical statements, if this dependence is taken into account within the formulae of estimation and prediction. However, for this purpose one has again to specify *assumptions* on the kind and extent of this dependence, shortly on its "regularity". Besides, also the usual assumption on the independence is an assumption on such an regularity.

Fore the sake of simplification of the representation in the following the points are denoted by $Q = (x, y, z)$ and the difference of the functional values is abbreviated by $G_P(Q_1, Q_2) = G_P(Q_1) - G_P(Q_2)$. According to the splitting formula (7.41) of G_P one obtains

$$\mathrm{E}\, G_P(Q_1, Q_2) = G_0(Q_1, Q_2) + \mathrm{E}\,(\epsilon(Q_1) - \epsilon(Q_2)) \qquad (7.42)$$

$$\mathrm{E}\, G_P^2(Q_1, Q_2) = G_0^2(Q_1, Q_2) + \mathrm{E}\,\epsilon^2(Q_1) + \mathrm{E}\,\epsilon^2(Q_2)$$
$$-2\mathrm{E}(\epsilon(Q_1)\epsilon(Q_2)) . \qquad (7.43)$$

Without any additional assumptions from these formulae for *one* realization no statistical conclusions can be drawn.

The *first* assumption concerns the *isotropy* of the field: The dependence of the random deviations depends only on the *distance* of the two points, i.e.

$$E(\epsilon\,(Q_1)\epsilon\,(Q_2)) = C\left(\sqrt{(x_1 - x_2)^2 + (y_1 - y_2)^2 + (z_1 - z_2)^2}\right), \quad (7.44)$$

where the function C is called the *covariance function* of the field. The validity of the isotropy assumption seems reasonable for *small* regions in application. Certain and different generalizations of the model are possible.

When considering two-dimensional cases in a plane region, then obviously the formulae above can be written also for points $Q = (x, y)$ and, particularly, isotropy and covariance are defined in the same manner. With respect to the third coordinate, say the depth z, also other regularities of dependence can be introduced and are used in application (see e.g. CRESSIE (1991)).

In geostatistics and in the case of isotropy instead of the covariance function the *variogram*, as it is called,

$$E\,G_P^2(Q_1, Q_2) = 2\gamma(Q, d) \qquad (7.45)$$

is considered, where $Q_1 = Q$ and $Q_2 = Q + d$ are replaced. The function γ is called the *semivariogram*. Semivariogram and covariance function can be computed from each other, according to (7.43) and (7.44), theoretically.

The problem, however, consists in that neither C nor γ is known. Their estimation is impossible, because only *one* realization is given. For a reasonable estimation of $g_0(Q)$ one needs information on the stochastically regularity of the deviations, i.e. on C or γ. So the estimation problem would be really *unsolvable*.

A *way out* is offered by the *intrinsic hypothesis*, as it is called: The covariance function, respective the semivariogram, does *not* depend on Q, but only on d. This assumption is fulfilled in practical situations again within local regions approximately. From this starting point there are *two* different approaches.

In the first approach one demands $E\epsilon\,(Q) = 0$ and assumes moreover that the field is homogeneous in a wide sense, i.e. among others $D^2\epsilon\,(Q) = E\epsilon^2(Q) = \sigma^2$. Then the trend part g_0 may have an arbitrary form. The intrinsic hypothesis applies here only to the covariance function.

With the second approach one demands $E(\epsilon\,(Q_1) - \epsilon\,(Q_2)) = 0$ and assumes moreover that the field has homogeneous increments, this means among others $D^2(\epsilon\,(Q_1) - \epsilon\,(Q_2))^2 = \tau^2$. Then $g_0(Q) = m$ must be constant. The intrinsic hypothesis applies here to the semivariogram.

With these assumptions estimations are possible. One obtains the *empirical variogram*, as it is called,

$$2\hat{\gamma}(h) = \frac{1}{n(h)} \sum_{i=1}^{n(h)} \left[g_P(Q_i) - g_P(Q_i + h)\right]^2 \qquad (7.46)$$

by adding up over all $n(h)$ couples of values at observation points Q_i, which have each the respective distance h.

The semivariogram can be *discontinuous* at the origin:

$$\gamma(0) = 0 \quad \text{but} \quad \lim_{h \to 0} \gamma(h) = C_0 > 0 . \qquad (7.47)$$

The magnetude of discontinuity C_0 is called *nugget-effect* and results from the existence of subdimensional structures (problem of the chosen scale) or by small additional measurement errors in the process of gathering the observation values. The typical behaviour of an empirical semivariogram is shown in Fig. 7.3. As a rule, a (parametric) setup is chosen for a representation of this semivariogram and its parameters are estimated from the given data. These estimates are then used for estimating the required quantities of the regionalized variables. A particular proceeding was called *Kriging*-technique according to a suggestion of MATHERON (1969), who intended by this to honour his friend KRIGE, being the first in applying this technique, in prospection and mining gold in South Africa.

First in a *trend analysis* the systematic part is splitted off. From the estimated deviations in the observation points the covariance function is estimated. With this estimate the trend estimate is improved and from this improved estimate the estimate of the covariance function is improved. This iteration procedure is continued until no essential improvements can be obtained.

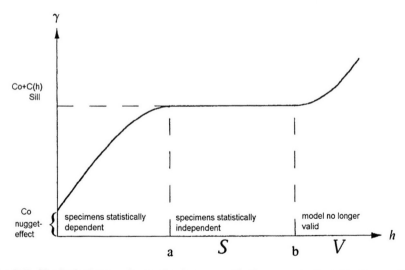

Fig. 7.3. Typical picture of a semivariogram with the nugget-effect C_0, the reach a, the stable region S and the region V, in which the intrinsic hypothesis is no longer valid

In the decades passed diverse variants, refinements and generalizations of this technique became known, which found their expressions in textbooks and software tools.

Because of the local character of the theory it suggests itself to use results from other regions, which are considered as similar for factual reasons to the situation in the present region. This invites to apply BAYESIAN methods (see PILZ/PLUCH/SPÖCK (2005), for a connection with considerations from fuzzy set theory see BANDEMER/GEBHARDT (2000)).

The above explanations should only serve to clarify the essence of the technique and to allow a reasonable assessment of the results obtained by it.

7.2 Fuzzy Evaluation of Functional Relationships

In the preceding section the uncertainty of the data, which served for an evaluation of the functional relationships, are modelled by a probabilistic background. They were interpreted as realizations of random variables, the properties of which were specified by a series of *assumptions* to be considered and checked in every case of application. In the third chapter two alternatives for specifying uncertainty were presented, which should be now confronted with the problem of evaluation of functional relationships.

As the simplest case an information situation is considered, in which the values of the explanatory variables are given *exactly* and for the values y_i of the response variable at the observation points $\mathbf{x}_i$ intervals of the form are specified

$$y_{ui} \leq y_i \leq y_{oi} \, . \tag{7.48}$$

A near at hand plausible proceeding would be the consideration of the set of all values of the parameter vector $\mathbf{a}$ of the functional relationship

$$y = \eta(\mathbf{x}, \mathbf{a}) \, , \tag{7.49}$$

for which the functional values lie between the corresponding lower and upper bound at each single observation point:

$$A = \{\mathbf{a} \mid \text{for all } i : y_{ui} \leq \eta(\mathbf{x}_i, \mathbf{a}) \leq y_{oi}\} \, . \tag{7.50}$$

This proceeding shows, however, two essential difficulties. On the one side the specification of the *crisp* bounds is not possible without a high arbitrariness, as already explained in Sect. 3.1. The solution set A could depend essentially on this specification (as in the case of the linear equation system with interval data presented in ALEFELD/MAYER (1995)). On the other side the computation of A in non-trivial setups might be very complicated and will be effected numerically only with an unjustifiably high expenditure.

Therefore, it is reasonable to go over immediately to evaluated intervals, i.e. to *fuzzy sets* (see Sect. 3.2). In using fuzzy sets for evaluation of functional relationships there are *two* different approaches.

In the one approach it is tried to generalize methods of mathematical statistics for *crisp* data to the case of fuzzy data in applying *extension principles* to the corresponding notions and expressions, as it was mentioned already in Subsect. 4.2.4 (see VIERTL (1996)).

The other approach transforms the *starting problem* and offers a solution by general methods from the theory of fuzzy sets.

7.2.1 Crisp Data Analysis as a Starting Point

Usually *crisp* data are given as points in a k-dimensional space. In the *explorative* phase of data investigation, if there is *not yet* given an expertly supported or a near at hand setup, data analysis starts with the arrangement of the points according to further properties, mostly with respect to their frequency in pre-assigned classes. This arrangement is connected with a search for outliers, the reliability and representativity of which is questioned, at least for the moment. Next it is tried, by suitable representations and transformations, to realize *structures* within the "point cloud", inter alia to obtain an idea for a setup. If the data space has a dimension higher than two or three, which is usually the case, then these transformations must be *projections* to allow a visual inspection at the screen. With series of such projections obeying a certain rule of generation, called *projection pursuit*, it is tried at the screen to find such projections, by which a special structure in the data becomes obvious. Such structures may be, e.g., a decomposition of the whole cloud into partial clouds or an arrangement of the bulk of the points along curves or surfaces. This technique was used first by FRIEDMAN/TUKEY (1974); the survey paper by HUBER (1985) can be recommended for a pregnant and coherent overview.

If the search is to be performed automatically, then the software must have some *measure of interestingness* given for every single found constellation, by which the computer "learns", which type of "structure" it should look for. This technique was generalized to fuzzy data (see BANDEMER/NÄTHER (1988a)), especially if the points are specified as of generalized "bean-type"

$$\mu(\mathbf{x}) = h(d(\mathbf{x}, \mathbf{x}_i)) , \tag{7.51}$$

where d is a distance function and h is a suitable monotonically increasing function (with respect to details see BANDEMER/NÄTHER (1988a)).

Most of the methods of classical multivariate analysis, as principal component analysis, discriminant analysis, and some methods of factor analysis, turn out to be partial cases of projection pursuit technique (with corresponding measures of interestingness).

A special case widespread in application is the *partial least squares technique*, as it is called, PLS for short. It goes back to WOLD (1985) and forms the essential element of his "soft modelling". This technique tries to evaluate (linear) functional relationships among non-observable (latent) variables from observable ones assumed to be dependent on those. Also this PLS-technique was adapted to fuzzy data (see BANDEMER/NÄTHER (1988b)). Taking into account data fuzziness in these procedures facilitates an essentially higher sensivity of cognition procedures, because functionally caused differences will be realized much better against data fuzziness, or possibly erroneously as functional interpreted differences may disappear in data fuzziness.

A simple method to come to an approximation for an assumed functional relationship in the crisp case is, as is well known, the strictly *connecting* of *neighbouring* observation points with each other in order to obtain an approximation at least in a local region. This method can be applied, obviously, also to fuzzy data.

To avoid a proliferation of notations only the simple two-dimensional special case $(x, y) \in I\!R^2$ is considered for the fuzzy data $\mathcal{Z}_i; i = 1, \ldots, n$ with the corresponding membership functions $\mu_i(x, y)$.

Since the supports of the single data $\text{supp}(\mathcal{Z}_i)$ can overlap, for a local representation first an aggregation principle must be chosen to obtain a unique aggregated datum. Usually the union is chosen for this purpose, mostly represented by the maximum.

Then a first and near at hand recommendation for a local approximation of the functional relationship of the response variable y and the explanatory variable x is to choose the *modal trace*

$$F_{\mathrm{mod}}(x) = \{y \,|\, y = \arg \sup_y \mu_Z(x, y)\}; \; x \in \{x \,|\, \sup_y \mu_Z(x, y) > 0\} \,, \quad (7.52)$$

i.e. for every single x the set of all y is considered having maximum membership to $\mathcal{Z}$ at x, naturally only within the set, where information is supplied by $\mathcal{Z}$ at all. This criterion can be interpreted in the sense of possibility theory (see Subsect. 5.2.2): as aiming at maximum possibility for the fitting values. In general the modal trace will present for some x several values of y. The trace will be neither unique nor continuous, but will show the behaviour of a natural range of mountains as can be seen in topographical maps. All these shortcomings will not be a disadvantage: for a first impression it suffices to realize occurring trends. Moreover, a premature smoothing and pressing to uniqueness and spreading into unobserved regions may alter the information inherent in the original data inadmissibly and can be misleading for the following analysis. If the trace branches out into several clearly separated ranges with respect to y, then one should inspect the original data for outliers, reconsider the specifications of the fuzzy sets, or, remind the possibility that the functional relationship can have an *implicit* form.

In the case of an implicit functional relationship the principle of modal trace can be modified. Interpreting now the values of the membership function μ_Z as values of height in a topographical map, then one may adopt a principle of *watersheds*, where only the ridges are chosen to support the approximation, which are watersheds of the fuzzy "mountains". Methods for calculating such watersheds can be found in BANDEMER/HULSCH/LEHMANN (1986).

A case study, including an application of this modal trace in a practical case to find an expertly supported setup, will be given within the next subsection.

7.2.2 Explorative Evaluation of Functional Relationships

In the preceding subsection the data were considered *locally*, i.e. each time in the immediate neighbourhood of each single datum. Now it is started aiming at a representation of the data by a *functional expression*. A functional relationship

$$y = \eta(\mathbf{x}, \mathbf{a}) \tag{7.53}$$

is given over the domain $\mathbf{x} \in B$. The parameter vector $\mathbf{a}$ may vary within the set A.

For a treatment of this *global* evaluation problem *two* different attitudes for a model understanding can be distinguished. In the *explorative* attitude, which is considered first, the data are taken as they are. Statements on the functional relationship use only these data and refer only to those parts of the aggregated datum, in which its membership function is positive, i.e. to $\text{supp}(\mathcal{Z})$. In the problem of approximation treated later on, assumptions are necessary, which allow the inclusion also of the domain *outside* the support.

For the explorative case one starts with the functional relationship (7.53) and tries to transfer the obtained (fuzzy) information contained in the fuzzy data $\mathcal{Z}_1, \ldots, \mathcal{Z}_n$ into the parameter set A of the functional relationship. The corresponding mapping

$$\mathcal{A} = V(\mathcal{Z}_1, \ldots, \mathcal{Z}_n) \tag{7.54}$$

is called *transfer principle*. The evaluation of every single function $\eta(\mathbf{x}, \mathbf{a})$ of the functional relationship (7.53) is a fuzzy evaluation of $\mathbf{a}$, i.e. by the membership function $\mu_A(\mathbf{a})$ of $\mathcal{A}$ from (7.54).

The mapping V consists of two operations: an *aggregating* one V_{agg}, which removes the dependence on the individuality of the data (the index i) and an *integrating* one V_{int}, which removes the dependence on the diversity of the x-points in the domain.

For mathematical convenience, theoretically as well as numerically, the two operations should be specified individually and applied one after another. The proposed transfer principle used differ from each other in the specification of the partial mappings and in the order of their application.

For aggregating the data one can first take into account for the functional relationship all those points, which belong to *at least* one of the data $\mathcal{Z}_i$. Then the data are to be *united*, e.g. by maximizing

$$\mu_Z(\mathbf{x}, y) = \max_i \mu_i(\mathbf{x}, y) \,, \tag{7.55}$$

as it had been practised in forming the modal trace.

If each single datum should be used at every point with equal rights, perhaps with different degrees of trustworthiness $\beta_i \in [0, 1]$, then the weighted mean is near at hand, i.e. computing an aggregated datum of another kind with the membership function

$$\mu_Z(\mathbf{x}, y) = \sum_{i=1}^{n} \beta_i \mu_i(\mathbf{x}, y) \,, \tag{7.56}$$

with suitable chosen $\beta_i's$, such that μ_Z remains in the unit interval.

If one takes for useful only such information that is contained in each single of the given data, then even an *intersection* of the data would make sense.

Finally, the transfer principle can be made more robust against outliers, if the aggregation is restricted to only a certain subset of the data (see BANDEMER/NÄTHER (1992)).

Now, one can apply an integrating operation V_{int} to the aggregated datum $\mathcal{Z}$. In general

$$\mu_{Z(\eta(.,\mathbf{a}))}(\mathbf{x}) = \mu_Z(\mathbf{x}, \mathbf{a}) \tag{7.57}$$

with $\mathbf{x} \in B$ represents the degree to which the graph $\{(\mathbf{x}, \eta(\mathbf{x}, \mathbf{a}))\}; \mathbf{x} \in B$ hits the aggregated datum $\mathcal{Z}$ in $\mathbf{x}$ for a given parameter value $\mathbf{a}$. Considered over B then $\mathcal{Z}(\eta(., \mathbf{a}))$ is a fuzzy set with the membership function (7.57). In this manner every integrating operation V_{int} is an evaluation principle for this set and hence for $\mathbf{a}$. For specifying a quite simple integrating operation one chooses some weight function w over B expressing some additional knowledge or desirable requirements. For example, by

$$w(\mathbf{x}) = w_0 \sup_y \mu_Z(\mathbf{x}, y) \tag{7.58}$$

every single $\mathbf{x}$ is weighted by the modal trace value of membership (7.52), and regions without any information by fuzzy sets are omitted.

Then

$$\mathcal{A}_1^* : \mu_{A_1^*}(\mathbf{a}) = \int_B \mu_Z(\mathbf{x}, \eta(\mathbf{x}, \mathbf{a})) w(\mathbf{x}) \mathrm{d}\mathbf{x} \tag{7.59}$$

is a *quantitative* evaluation and together with the chosen aggregation principle V_{agg} it is a useful tranfer principle. If $w(x)$ is interpreted as some probability density, then (7.59) gets the character of an expectation value. This tempted to use the misleading name "transfer principle of expected cardinality" in the first publication of this principle (see BANDEMER (1985)).

Obviously, a probabilistic interpretation is possible, but not necessary (see BANDEMER/NÄTHER (1992)).

Finally, the membership function of $\mathcal{Z}(\eta(.,\mathbf{a}))$ can also be interpreted as a possibility distribution leading to

$$\mathcal{A}_2^* : \mu_{A_2^*}(\mathbf{a}) = \sup_{\mathbf{x}} \mu_Z(\mathbf{x}, \eta(\mathbf{x}, \mathbf{a})) \ . \tag{7.60}$$

The given examples for both the partial operations may suffice to give an idea of their aim and form. Obviously, each aggregating operation $\mathcal{Z} = V_{agg}(\mathcal{Z}_1, \ldots, \mathcal{Z}_n)$ can be combined with each integration operation $\mathcal{A}^* = V_{int}(\mathcal{Z})$, if this combination makes sense in the practical situation.

Now, the *opposite* order of applying the operations is considered, i.e. $\mathcal{A}_i = V_{int}(\mathcal{Z}_i)$ and $\mathcal{A}^* = V_{agg}(\mathcal{A}_1, \ldots, \mathcal{A}_n)$.

A first suggestion for V_{int} is motivated by the extension principle and leads to the membership function

$$\mu_{A_i,1}(\mathbf{a}) = \sup_{(\mathbf{x}, y): y = \eta(\mathbf{x}, \mathbf{a})} \mu_i(\mathbf{x}, y) = \sup_{\mathbf{x}} \mu_i(\mathbf{x}, \eta(\mathbf{x}, \mathbf{a})) \ . \tag{7.61}$$

The membership function for $\mathcal{A}_i$ can be interpreted as the uncertainty induced by the fuzzy datum $\mathcal{Z}_i$ to the parameter set A, possibly as a sort of possibility distribution. The same operation is obtained, if the usual statement on the validity of a relation in a given set is fuzzified (see BANDE-MER/SCHMERLING (1985) and BANDEMER/NÄTHER (1992)). Obviously, the integral (7.59) can be computed also for each single i separately leading to

$$\mu_{A_i,2}(\mathbf{a}) = \int_B \mu_i(\mathbf{x}, \eta(\mathbf{x}, \mathbf{a})) w_i(\mathbf{x}) d\mathbf{x} \ , \tag{7.62}$$

where the weight function may vary from datum to datum according to the practical context.

The fuzzy sets $\mathcal{A}_i$ can now be aggregated by some chosen operation V_{agg}. The *intersection*, e.g. represented by the minimum,

$$\mu_{A_3^*}(\mathbf{a}) = \min_i \mu_{A_i}(\mathbf{a}) \ , \tag{7.63}$$

could be explained as the degree to which at least each single fuzzy datum $\mathcal{Z}_i$ contains a point of the functional relationship $\eta(\mathbf{x}, \mathbf{a})$. Hence $\mathcal{A}_3^*$ was at first called also the "joint grade of validity of the functional relationship" with respect to the given data (see BANDEMER/SCHMERLING (1985)). Naturally, one can choose for minimizing only a suitable part of the data, to make the procedure robust against outliers.

Another possibility is again an aggregation by the weighted mean. This is preferred, if the data are obtained by frequency analysis.

With respect to other principles and, generally, with regard to their special mathematical properties and distinctions it is referred to BANDE-MER/NÄTHER (1992).

To give an idea how these proposals do work in a practical context in evaluating a functional relationship from *fuzzy* data, in the following a practical application with real data is sketched. A detailed presentation, including the original numerical data, can be found in BANDEMER/KRAUT/VOGT (1988) and, each time shortened, in BANDEMER/KRAUT (1990) and BANDEMER/NÄTHER (1992).

One of the most common procedures in testing of material is the measurement of its hardness. A usual method to measure VICKER's hardness consists in the following performance. A regular quadrangular pyramid made of diamond is pressed with a certain power onto the surface of the specimen. When the pyramid is removed the specimen shows a remaining impress of quadrangular shape. Hardness h is then defined as the quotient of the pressing power p and the area of the impress surface, say s, on which the power worked

$$h(p, s) = \frac{p}{s} \, . \tag{7.64}$$

Let d denote the length of the diagonal of the square base of the pyramid, then one obtains the area of the impress by

$$s = \frac{d^2}{c_0} \, , \tag{7.65}$$

where c_0 depends on the face vertex angle of the pyramid.

The following problem of fuzzy data analysis was considered. A rectangular solid specimen was subjected to a hardening treatment onto one of its faces. Then the specimen was cut up orthogonally to this treated face. The inner plane produced in this manner was covered by a grid of points, at each single point of which VICKER's hardness was measured. From the results of these measurements a *functional relationship* should be evaluated connecting the *hardness* and the distance to the border, where the hardening treatment was applied, shortly called the *depth*. This functional relationship should then be used to optimize and control the hardening process and hence should have a form as simple as possible and expertly supportable.

The specimen was very small, caused by the intended application of the results. (As is well known, the transfer of results to other magnitudes is problematic. Laboratory micro results may loose their sense frequently when being transferred to a production scale, whereas, on the other side, results from a usual experimental department need no longer be valid in a micro region.) Moreover, for a high resolving power with respect to position, both, pressing power and impress, must be very small. The observation must be performed by presenting the specimen plane to a microscope and by enlarging the picture by means of an image processing equipment. The result of the observation was presented as a grey-tone picture of the impress grid on the screen. The diagonal of each single impress was to be measured to obtain the corresponding local hardness.

If the effect of the pressing procedure is considered more precisely, one realizes that the impress in the material is *not* an *exact* copy of the pressing pyramid, because material was pressed *out* of the produced cavern and formed an *embarkment* around the impress changing its shape and preventing an *exact* measurement of the diagonal. Moreover, the *two-dimensional* image of the *three-dimensional* impress on the screen acts as an another source of impreciseness, which cannot be controlled by the observer, but influences the result perhaps essentially. The human eye as a measuring tool, the light-optical lower bound of resolving power, wavelength of light, adjusting conditions at the microscope, scanning quality of the television equipment and the internal automatical control of brightness and contrast are mentioned in this respect. Note that a sharpening of edges in the grey-tone picture by means of mathematical morphology might raise precision only virtually, because it induces some arbitrariness by the observer.

Moreover, the coordinates of the points, where hardness was measured on the specimen, are likewise subject to inaccuracy and uncertainty. The impress has a *finite* extent and thus measures hardness *in a certain neighbourhood* of the putting down point of the top of the pyramid. Additionally, the screen shows always only a *part* of the plane segment under investigation and turning from one segment to the next one causes increasing inaccuracy in specifying the respective depth.

Hence both, hardness and depth, were modelled by fuzzy sets.

The grey-tone pictures of the impresses were interpreted as fuzzy sets $\mathcal{G}_i$ in the plane $I\!\!R^2$, where the shades should reflect the vagueness of the experimental result. The diagonals of these fuzzy squares $\mathcal{G}_i$ had the same directions each as the respective axes of the chosen coordinate system. For measuring the lengths of the diagonals the *fuzzy regions* $\mathcal{G}_i$ were transformed into their corresponding *fuzzy contours* $\mathcal{C}_i$

$$\mathcal{C}_i : \mu_{C_i}(x,y) = 2\min\{\mu_{G_i}(x,y), 1 - \mu_{G_i}(x,y)\} . \tag{7.66}$$

The membership function of a fuzzy region $\mathcal{G}$ indicates the degree of membership of a point to the region itself, but not to the *border* of this region, the *contour* C. Formula (7.66) generalizes the usual conception of the border for *crisp* regions that the border should belong to the closed region and simultaneously to the closed complement of the region. For details and general proposals see BANDEMER/KRAUT (1988) and BANDEMER/NÄTHER (1992).

Let y_{0i} be the coordinate of the horizontal diagonal of $\mathcal{C}_i$. The correponding membership function $\mu_{C_i}(x, y_{0i})$ splitted into two disjoint curves, which could be interpreted as the membership functions of two fuzzy numbers $\mathcal{M}_{li}$ and $\mathcal{M}_{ri}$. They indicated the positions of the *left* and *right* end, respectively, of the diagonal on the x-axis, which denotes here the depth. Their difference

$$\mathcal{D}_i = \mathcal{M}_{ri} \ominus \mathcal{M}_{li} \tag{7.67}$$

defines the *fuzzy length* of the diagonal. An inspection of the α-cuts of $\mathcal{D}_i$ by the grey-tone levels of the picture on the screen showed that an approximation of the membership function of $\mathcal{M}$ by linear reference functions was appropriate. In this manner also $\mathcal{D}$ is of such a simple form, namely

$$\mathcal{D}_i = \left\langle d_i; l(d_i), r(d_i) \right\rangle ; \tag{7.68}$$

where d_i is the core of $\mathcal{D}_i$, and $l(d_i)$ and $r(d_i)$ are the left and right spread, respectively, of the fuzzy number $\mathcal{D}_i$ (see (3.49)). Since $\mathcal{D}_i$ belongs to the point x_i where the top of the pyramid touched the plane, in (7.68) the dependence of the fuzzy diagonal length on the depth of the corresponding observation is indicated explicitly.

For convenience the constant c_0 is omitted in the hardness formula (7.65) in the following. The *fuzzy hardness* $\mathcal{H}(x)$ at the *crisp* depth x is computed from the *fuzzy diagonal length* $\mathcal{D}_i$ according to the extension principle

$$\mu_{H(x|x_i)}(h) = \sup_{v:h=p/v^2} \mu_{D_i}(v) = \mu_{D_i}((p/h)^{1/2}) . \tag{7.69}$$

For modelling the fuzziness of the specification of depth $\mathcal{X}(x_i)$ also a fuzzy number with a triangular membership function is assumed, which might be symmetric, since a preference of one of the directions for the fuzziness did not seem reasonable:

$$\mathcal{X}(x_i) = \left\langle x_i; c(x_i), c(x_i) \right\rangle . \tag{7.70}$$

Because there is no reason to assume an interaction between the measurement of hardness and the specification of depth, the two fuzzy sets are combined by the common Cartesian product, here represented by the minimum operator:

$$\mu_{H(X(x_i))}(h, x) = \min\{\mu_{H(x|x_i)}(h), \mu_{X(x_i)}(x)\} . \tag{7.71}$$

Considered as a surface over the (x, h)-plane this function looks like a rectangular tent with four curved roof edges from the top. With respect to the analytical form of the function see the paper by BANDEMER/KRAUT/VOGT (1988).

For evaluating the functional relationship between hardness and depth the fuzzy observations $\mathcal{H}(X(x_i))$ at the different observation points are aggregated by union, represented by the maximizing operator

$$\mu_H(h, x) = \max_i \mu_{H(X(x_i))}(h, x) . \tag{7.72}$$

For a first impression on the possible functional relationship between h and x the *modal trace* of $\mu_H(h, x)$ is considered, i.e.

$$H_F(x) = \{(h, x) : \mu_H(h, x) = \sup_u \mu_H(u, x) > 0\} . \tag{7.73}$$

It is marked in Fig. 7.4 by a lighter grey-tone within the grey band. This modal trace suggested the functional relation

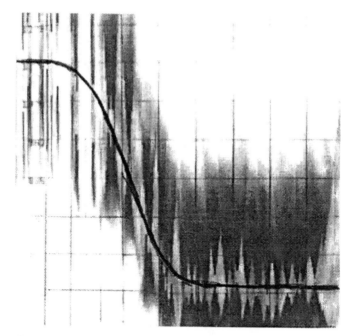

Fig. 7.4. Result of the fuzzy data analysis for the hardness example: the fuzzy functional relationship as a grey-tone picture and the "best" crisp estimating curve through the modal trace

$$h(x; a, b, \nu, q) = a + b \exp \left\{ - \left(\frac{x}{\nu} \right)^q \right\} , \qquad (7.74)$$

which allowed an expertly supported interpretation. The parameter a represents the hardness of the kernel of the material (not influenced by the hardening treatment), b is the maximum hardness increase at the frontal surface, and ν and q explain, where and how fast hardness decreases with increasing depth. With this setup for each single interesting quadruplet (a, b, ν, q) the integral principle according to (7.59) was applied, where the characteristic function of the support $S = \text{supp} H_F(x)$ was used as a weighting function $w(x)$. This led to a fuzzy set $\mathcal{A}^*$ over the four-dimensional parameter space with the membership function

$$\mu_{A^*}(a, b, \nu, q) = \int_S \mu_H(h(x; a, b, \nu, q), x) \mathrm{d}x / \int_S \mathrm{d}x . \qquad (7.75)$$

This approach was also technologically the simplest one, because it could be effected immediately on an image processing equipment, which supplied integration results by summing up the grey-tone values with sufficiently numerical precision. Because the result was practically exceptionally convincing, no other transfer principle was tried. A short and simple optimizing procedure to find a function with highest membership resulted in a curve marked in Fig. 7.4. The numerical results are recorded in BANDEMER/KRAUT/VOGT (1988).

7.2.3 Evaluation with Additional Assumptions

Up to now the fuzzy data were treated only in an exploratory sense, outside the supports of the data no information was expected or used. This led to the recommendation to consider only that part of B, where the united datum Z has positive membership values. It can happen, however, that there is *no* function of the functional relationship, which hits *all* the fuzzy data simultaneously. In this case and by a union according to the minimizing principle it led to a fuzzy parameter set $\mathcal{A}^*$, which does *not* contain any element, although a lot of graphs of functions of the functional relationship run through the data "cloud".

Hence there are some proposals to overcome this unwanted property taking into account *neighbourhoods* of the functions and/or of the data. The crisp or fuzzy approximating parameter for the desired approximation is then determined by a given optimizing procedure. In the following three proposals of such *approximation principles* will be sketched.

In the first option, the *transfer principle for belts* is considered, by which the function $\eta(\mathbf{x}, \mathbf{a})$ of the functional relationship is replaced by a whole functional "belt"

$$\eta(\mathbf{x}, \mathbf{a}) + \Delta; \quad \Delta \in I\!\!R^1 . \tag{7.76}$$

For each fixed $\Delta_0 > 0$ this specifies a neighbourhood for $\eta(\mathbf{x}, \mathbf{a})$ with $|\Delta| \leq \Delta_0$. Introducing (7.76) into some chosen tranfer principle one obtains a fuzzy set over $A \times I\!\!R^1$. The membership function $\mu(\mathbf{a}, \Delta)$ of which can now be maximized with respect to $|\Delta| \leq \Delta_0$. More details for this approach see BANDEMER/NÄTHER (1992).

Another approach of "broadening" the function $\eta(\mathbf{x}, \mathbf{a})$ would be its replacing by a fuzzy function $\mathcal{H}(\mathbf{x}, \mathbf{a})$, for which η represents the core function (with $\mu = 1$). Then the *individual* distance of each single datum from this *fuzzy* functional relationship is considered, i.e.

$$d(Z_i, \mathcal{H}(\mathbf{x}, \mathbf{a})) , \tag{7.77}$$

where d is here a suitable distance function between fuzzy sets.

The general principle consists in that the distances are aggregated with respect to i and parameter values $\mathbf{a} \in A$ are to be computed, for which the aggregated distance is minimum.

Examples for such a proceeding can be found in DIAMOND (1988), ALBRECHT (1992), CELMINS (1987), on which it is reported in BANDEMER/NÄTHER (1992).

Finally, the approach by TANAKA is mentioned (see TANAKA/UEJIMA/ASAI (1982), TANAKA/WATADA (1988)), which allows fuzziness only in y-direction represented by fuzzy numbers and which considers only special (linear) relationships (see also BANDEMER/NÄTHER (1992)). This approach has already gained a certain spreading and adapting in application.

All these approaches introduce each an optimizing criterion, which could be justified by a *decision problem* in the background and by assumptions on the regularity of the *data genesis*. *Without such a motivation* these criteria are arbitrary restrictions of the information transfer with the aim of only an increase of *definiteness* of the respective result, which but decreases the value of the obtained statements really.

7.2.4 Inference with Fuzzy Parameter Values

In mathematical statistics the obtained estimates of the parameters in a regression setup are used for different problems of inference, e.g. for predicting the response at further observation points or for discriminating among competing setups.

In fuzzy data analysis for evaluating functional relationships one obtains a *fuzzy* parameter set $\mathcal{A}^*$, e.g. by some extension principle. With such a set a parameterfuzzy functional relationship over B

$$y = \eta(\mathbf{x}, \mathcal{A}^*) \tag{7.78}$$

can be specified, which will form the basis of a correponding *fuzzy inference*. The star at $\mathcal{A}$ is suppressed in the following, since the origin of the fuzzy set is not essential for the fuzzy inference itself. For the practical interpretation, however, and the assessment of the *inference results* this origin plays its role, obviously. If $\mathcal{A}$ would be, e.g., a fuzzy specification by experts, then the credibility of the inference results would be not higher than that of this specification.

The usual inference is effected by the extension principle. Especially, at the *crisp* point x_0 one obtains as an *interpolation* the *fuzzy* functional value $\mathcal{y}$

$$\mu_Y(y; x_0) = \sup_{\mathbf{a}: y = \eta(x_0, \mathbf{a})} \mu_A(\mathbf{a}) \tag{7.79}$$

and at the *fuzzy* point $\mathcal{X}_0$

$$\mu_Y(y, \mathcal{X}_0) = \sup_{\mathbf{a}: y = \eta(x, \mathbf{a})} \min\{\mu_A(\mathbf{a}), \mu_{X_0}(x)\} . \tag{7.80}$$

Naturally, these formulae can be used for *extrapolation* as well, but with due caution, since extrapolation includes the silent assumption that the model, i.e. the form of the relationship, is valid also in the point under consideration.

This assumption may fail outside the region, where the observations were performed to find the fuzzy parameter set.

In contrast with the difficulties found in statistical inference the problem of *calibration* is now symmetric to that of interpolation. For the given *crisp* value y_0 the *fuzzy* argument value $\mathcal{X}$ is *calibrated* by

$$\mu_X(x; y_0) = \sup_{a:y_0=\eta(x,\mathbf{a})} \mu_A(\mathbf{a}) , \qquad (7.81)$$

and for the *fuzzy* value $\mathcal{Y}_0$ one computes the *fuzzy* calibrated argument value $\mathcal{X}$ by

$$\mu_X(x, \mathcal{Y}_0) = \sup_{(\mathbf{a},y):y=\eta(x,\mathbf{a})} \min\{\mu_A(\mathbf{a}), \mu_{Y_0}(y)\} . \qquad (7.82)$$

Applications of these formulae in a practical context of chemometrics can be found in OTTO/BANDEMER (1986), (1988a), (1988b).

Another interesting problem of inference consists in the combination of fuzzily expressed information on the parameter **a** from different sources. Let be, e.g., $\mathcal{A}_P$ a fuzzy set specifying a-priori knowledge on **a**, and $\mathcal{A}_E$ be the result of an estimation of **a** according to a transfer principle.

In a first step one could multiply both the membership functions each with a respective factor evaluating the contained information with respect to its trustworthiness, from an objective or from a subjective point of view. Then, in a second step, one can choose a connection to combine the two sets. Finally, in a third step, the obtained result can be renormalized in a certain sense. Particularly, if the two sets are to be used *with equal rights*, combined by *intersection* expressed by the *minimum*, and renormalized by the *supremum*, then the result is

$$\mu_A(\mathbf{a}|\mathcal{Z}) = \frac{\min\{\mu_P(\mathbf{a}), \mu_E(\mathbf{a})\}}{\sup_{\mathbf{b}\in\mathcal{A}} \min\{\mu_P(\mathbf{b}), \mu_E(\mathbf{b})\}} ; \qquad (7.83)$$

where $|\mathcal{Z}$ means that the data $\mathcal{Z}_1, \ldots, \mathcal{Z}_n$ are taken into account via $\mathcal{A}_E$. This is a *fuzzy analogue* of the well-known BAYES' theorem (see (4.38)). If the membership functions are interpreted as possibility distributions (see (5.9)), then in (7.83) a possibilistic a-priori distribution is coupled with a current possibility distribution yielding an a-posteriori distribution.

The problem of *model discrimination* occurs, if *several* functional relationships

$$\eta_1(\mathbf{x}, \mathbf{a}_1), \ldots, \eta_r(\mathbf{x}, \mathbf{a}_r); \quad \mathbf{a}_j \in A_j; \quad j = 1, \ldots, r \qquad (7.84)$$

compete to be used, motivated each single one either by an expertly supported idea or, merely, by the demand for a good approximation with a number of parameters as small as possible. The choice should be made with respect to the given data $\mathcal{Z}_1, \ldots, \mathcal{Z}_n$. A near at hand procedure consists in an evaluation of the parameters in each single given functional relationship, according to the same tranfer principle, and in an assessment of the respective fitting of the data by the obtained parameterfuzzy functional relationships. As a measure of good fitting a *fuzziness measure* would be useful. (A fuzziness measure is a set function, which evaluates every single set globally with respect to its fuzziness, e.g. to its deviation from the type of crisp sets. For details see BANDEMER/GOTTWALD (1995).) The chosen fuzziness measure reflects the desirable kind of goodness according to the user's ideas. As an example the transfer principle with $V_{int} = \sup_x$ and $V_{agg} = \min_i$ will be considered. Then

$$\sup_{\mathbf{a}_j \in A_j} \min_i \sup_x \mu_i(x, \eta_j(x, \mathbf{a}_j)) \tag{7.85}$$

would be a useful criterion for the selection of η_j. The value in (7.85) expresses, for each single j, namely the highest membership value that the graph of the respective functional relationship can reach for *all* the fuzzy data *simultaneously*. If there is only one datum, which is *not* met by this graph, the corresponding membership function vanishes. Therefore, the principle is a good one for model discrimination. However, the decision bases each time on only one (the highest) value; hence one should take into consideration also the whole membership functions of the single $\mathcal{A}_j^*$, especially if the decision is discussed.

8

Outlook and Conclusions

The "classical" treatment of a problem in application of mathematics consists, as is well known, in its idealization to a mathematical task, e.g., to a system of equations, or to a differential equation, or to an eigenvalue problem. This mathematical task is then solved exactly, and, if possible, in a closed analytical form. The quality of the solution is assessed by comparison with actuality. An example for such a proceeding was cited in Chap. 1 by the mathematical treatment of the flat rolling process.

However, such text-book applications are rather seldom. Frequently, the structure of the approximating task obtained by idealization and abstraction reflects the practical problem only essentially, and the data supplied by the practical situation are rather rough.

This problematic concerned some mathematicians open-minded to application already for many years. As a typical example from the point of view of interval mathematics the paper by NUDING (1975) may be cited. By a similar intention also this book was originated.

So, on the one side, interpolation and approximation are considered on different assumptions on the background and, on the other side, the effect of fuzziness, variability, and vagueness of data is investigated in simple tasks of qualitative and quantitative data analysis.

As is well known, such approximation approaches with data, possibly afflicted with uncertainty, are inserted as blocks in software packages for mathematical statistics and data analysis, for methods of finite elements, and for genetic algorithms, to cite only the best known fields. The development here goes to always more extensive, more elaborated, and allegedly more effective packages with always more comfortable output scenarios (tools, workbenches).

As a rule, the user is only said *what* the package is *able to solve*. *What* mathematical procedures *exactly* and in what manner were implemented, is frequently left in darkness. But even if all this would be explained, the common user could assess the procedure only seldom.

Moreover, different packages and techniques for solving the same problem are offered, e.g. from data analysis and from neural networks for cluster analysis. Because for each approach anyway there are still several competing software packages, the search for a suitable procedure can lead with the user to a "blackout by information overflow". In thus manner the final choice of a procedure is widely at random and arbitrarily.

A phenomenon interesting for mathematicians and regrettable for users is that the same mathematical procedures, algorithms, and techniques are "invented" again and again under new names each time. The present book may show, e.g., how the approximation with the method of least squares, considered against different backgrounds, leads to various interpretations of the same mathematical principle. A veritable Babel of languages would appear presumably, when the different mathematical terms, obtained by this principle, are listed accompanied by all the names given each single term in the various connections of backgrounds and application fields.

In the present book a lot of space is devoted to the question, how impreciseness, variability, and vagueness in the data influence the result. This problem does not play, as a rule, any role in software packages, if they are not concerned explicitly with even the handling of such data. It is, however, of importance in *every case*, in order to supply the user with an impression on preciseness and reliability of the results obtained. Moreover, it is relevant to the procedure itself; a highly precise computation (e.g. with the method of finite elements) with boundary values and perhaps also with coefficients of low precision does not make sense usually.

In the opinion of the author the development in this field runs in the *wrong* direction at present. Not always more extensive collections of methods with always "more precise" numerical procedures on always faster computers can be the aim of the development - it makes mathematics a sort of mysticism and shows the known "airport effect": Whereas the flying time from one airport to another one was shortened again and again with faster jets, the time necessary for the journey to the airport, from starting at home until the take-off, grew longer and longer, simultaneously. This means, in analogy, for the use of computing equipment that, before the task can be handled by the computer, one has to find first a suitable procedure within a package, then to understand the version found, and to interpret the given programme parameters meaningfully, and finally to specify them in a correct manner. Therefore the decision "all options" is so popular, but it does also not lead, as a rule, to an adequate treatment of the practical problem. It would be more reasonable, if the growing efficiency of computer equipment would be used to make software "intelligent", i.e. that computation is performed really reasonably: to create a *reasonable computing* as it was formulated in the subtitle of the book prognostically and programmatically.

To this end one should *first* think over what *aim* is actually pursued by the solution of the practically given problem. From this the level of detail is derived for the mathematical model necessary for that problem. Examples are given particularly in the Chaps. 2, 6, and 7.

Then the *quality* of the available data should be inspected to get an idea with respect to the background to be chosen.

From both considerations results then the choice of the solving procedure *and* the manner of interpretation of the result obtained.

Software packages, which should support the users in this process, should have a considerably other structure than the available ones at present. This requires changing in views with the mathematician, who has devoted himself to application, as well as with the software designer, who has to see over and above his collection package of methods.

A mathematician, who wishes to *apply* mathematics *really*, must first put aside his esoteric aura, in many places still very popular, and think oneself as a toolmaker for brain tools in a service position. For him mathematics is no longer organized only in disciplines (as algebra, analysis, stochastics, and numerics, to cite only the most common), but according to *abstraction principles*: what is considered as *essential* in the practical problem; what *mathematical structure* is suitable with the given aim to represent this in the best manner. Up to now the consultative mathematician recommends, as a rule, models and methods from his special discipline. One has to escape from this *hammer-nail-strategy* to get mathematics effective for application in its full ability.

In the next step fuzziness and impreciseness, variability and dependency must be considered, whether and how they should be specified and modelled, in order to choose a mathematical solving procedure and to be able for a reasonable interpretation of the result.

The software must *support* the user in doing so. It would be already a first step in this direction, if these problems would be picked out as themes and the user would be asked questions in this respect. These questions should be formulated by experienced users, because otherwise the questions would be felt as packed on top and hence ignored. Moreover, the user should be kept informed on the consequences of his decisions and hence put into the position to interpret the results obtained in this way also with respect to their precision, reliability and relevance. So, the input data should always be analysed with respect to their character and their numerical precision by the programme (with the user) and the parameters of the procedure should always be reasonably adjusted and adapted with respect to numerical precision (also to the internal precision of the procedure).

Such a new attitude has, naturally, also consequences for mathematical *instruction* of students from applied fields of knowledge outside mathematics. The aspect of modelling should take precedence over that of teaching procedures with all details. The reproduction of text-book examples with pseudo-exact data should be replaced by the discussion of modelling methods starting from practically typical situations. In such a course *each* discipline of mathematics should get the opportunity to present its structures of thinking as well as its possibilities of modelling.

Also for modelling of the practically relevant data impreciseness a training with standard situations should be offered. Here the starting point should be the demonstration of its specification and treatment.

Both these aspects should then be completed by practical work with computers. In doing so the details of the software packages used can be left in the background, because the software changes, as experience shows, rather frequently in its outfit.

Naturally it is clear that this type of instruction in mathematics of students from applied fields needs also another type of "teacher": a mathematician with knowledge from many disciplines and with practical experience - currently a rather rare species at universities.

If the present book can help only a little bit to lift mathematics for users out of its role as a source of heuristics, eclectic compilations, and recipes, and to turn it again into its role as a science of rationalization in thinking on the basis of logically founded models and procedures, then the book has fulfilled its purpose completely as planned by the author.

References

Agresti, A.: Categorial Data Analysis. John Wiley & Sons, New York (1990).

Albrecht, M.: Approximation of functional relationships to fuzzy observations. Fuzzy Sets and Systems 49 (1992) 301–305.

Alefeld, G.; Herzberger, J.: Introduction to interval computations. Academic Press, New York (1983).

Alefeld, G.; Mayer, G.: On the symmetric and unsymmetric solution set of interval systems. SIAM J. Matrix Anal. Appl. 16 (1995) 1223–1240.

Anderberg, M. R.: Cluster Analysis for Applications. Academic Press, New York (1973).

Ball, G. H.; Hall, D. J.: ISODATA: A novel method of data analysis and pattern classification. Techn. Report, Stanford Research Inst (1965).

Bandemer, H.: Evaluating explicit functional relationships from fuzzy observations. Fuzzy Sets and Systems 16 (1985), 41–52.

Bandemer, H.; Gebhardt, A.: Bayesian fuzzy kriging. Fuzzy Sets and Systems 112 (2000), 405–410.

Bandemer, H.; Gottwald, S.: Fuzzy Sets, Fuzzy Logic, Fuzzy Methods, with Applications. John Wiley & Sons, Chichester (1995).

Bandemer, H.; Hulsch, F.; Lehmann, A.: A watershed algorithm adapted to functions on grids. EIK 22 (1986) 553–564.

Bandemer, H.; Kraut, A.: On a fuzzy-theory-based computer-aided particle shape description. Fuzzy Sets and Systems 27 (1988) 105–113.

Bandemer, H.; Kraut, A.: A case study on modelling impreciseness and vagueness of observations to evaluate a functional relationship. In: Progress in Fuzzy Sets and Systems (Janko, W.; Roubens, M.; Zimmermann, H.-J., Eds.), Kluwer Academic Publ., Dordrecht (1990) 7–21.

Bandemer, H.; Kraut, A.; Näther, W.: On basic notions of fuzzy set theory and some ideas for their application in image processing. In: Geometrical Problems of Image Processing (Hübler et al. eds.), Akademie-Verlag, Berlin (1989) 153–164.

Bandemer, H.; Kraut, A.; Vogt, F.: Evaluation of hardness curves at thin surface layers–A case study on using fuzzy observations. In: Some Applications on Fuzzy Set Theory in Data Analysis, (Bandemer, H. Ed.), Deutscher Verlag für Grundstoffindustrie, Leipzig (1988) 9–26.

Bandemer, H.; Lorenz, M.: Evaluating a functional relationship from picture data of a chemical diffusion process – a case study -. Fuzzy Sets and Systems 93 (1998), 29–35.

Bandemer, H.; Näther, W.: Theorie und Anwendung der optimalen Versuchsplanung II. Handbuch zur Anwendung. Akademie-Verlag, Berlin (1980).

Bandemer, H.; Näther, W.: Fuzzy projection pursuits. Fuzzy Sets and Systems 27 (1988a) 141–147.

Bandemer, H.; Näther, W.: Fuzzy analogues to partial-least-squares techniques in multivatiate data analysis. In: Some Applications of Fuzzy Set Theory in Data Analysis, (Bandemer, H. Ed.), Deutscher Verlag für Grundstoffindustrie, Leipzig (1988b) 62–77.

Bandemer, H.; Näther, W.: Fuzzy Data Analysis. Kluwer, Dordrecht (1992).

Bandemer, H.; Pilz, J.; Fellenberg, B.: Integral geometric prior distributions for Bayesian regression with bounded response. Statistics 17 (1986) 323–335.

Bandemer, H.; Schmerling, S.: Evaluating explicit functional relationships by fuzzifying the statement of its satisfying. Biom. J. 27 (1985) 149–157.

Bandemer, H.; Schulze, U.: On estimation in multiphase regression models with several regressors using prior knowledge. Statistics 16 (1985), 3–13.

Berger, J. O.: Statistical Decision Theory and Bayesian Analysis. Springer, New York (1985).

Bezdek, J. C.: Pattern Recognition with Fuzzy Objective Algorithms. Plenum Press, New York (1981).

Bonner, R. E.: On some clustering techniques. IBM Journal 22 (1964) 22–32.

Boole, G.: The investigation of the laws of thought on which are founded the mathematical theories of logic and probabilities. McMillan (1854), Dover reprint 1958.

Box, G. E. P.; Jenkins, G. M.: Time Series Analysis. Forecasting and Control. Holden-Day, San Francisco (1970).

Box, G. E. P.; Wilson, K. B.: On the experimental attainment of optimum conditions. J. Roy. Statist. Soc., Ser. B. 13 (1951) 1–45.

Cantor, G.: Über unendliche, lineare Punktmannigfaltigkeiten (Arbeiten zur Mengenlehre aus den Jahren 1872–1884) Herausgegeben und kommentiert von G. Asser Teubner-Archiv zur Mathematik, Bd. 2 Teubner, Leipzig (1984).

Celmins, A.: Least squares model fitting to fuzzy vector data. Fuzzy Sets and Systems 22 (1987) 245–269.

Cochran, W. G.: Sampling Techniques. John Wiley & Sons, New York (1957).

Courant, R.; Friedrichs, K.; Lewy, H.: Über die partiellen Differenzengleichungen der mathematischen Physik. Math. Annalen 100 (1928) 32–74.

Cressie, N.: Statistics for Spatial Data. John Wiley & Sons, New York (1991).

deFinetti, B.: Theory of Probability. Vol I and II. John Wiley, New York (1937).

Dempster, A. P.: Upper and lower probabilities induced by a multivalued mapping. Ann. Math. Stat. 38 (1967) 325–329.

Diamond, Ph.: Fuzzy least squares. Inf. Sciences 46 (1988) 141–157.

Dubois, D.; Prade, H.: Operations with fuzzy numbers. Intern. J. System Sci. 9 (1978) 613–626.

Dubois, D.; Prade, H.: Fuzzy Sets and Systems. Theory and Applications. Academic Press, New York (1980).

Dubois, D.; Prade, H.: Possibility Theory. An Approach to Computerized Processing of Uncertainty. Plenum Press, New York (1988).

Ferguson, T. S.: Mathematical Statistics – a decision theory approach. Academic Press, New York (1967).

Filzmoser, P.; Viertl, R.: Testing hypotheses with fuzzy data. The fuzzy p-value. Metrika 59 (2004), 21–29.

Friedman, J. H.; Stuetzle, W.: Projection pursuit regression. J. Amer. Statist. Assoc. 76 (1981) 817–823.

Friedman, J. H.; Tukey, J. W.: A projection pursuit algorithm for exploratory data analysis. IEEE Trans. Comput. 23 (1974) 881–889.

Goetscherian, V.: From binary to grey-tone image processing using fuzzy logic concepts. Pattern recognition 12 (1980) 7–15.

Goldberg, D. E.: Genetic Algorithms in Search, Optimization, and Machine Learning. Addison-Wesley Publ. Comp., Reading (1989).

Goodman, I. R.; Nguyen, H. T.: Uncertainty Models for Knowledge-Based Systems. North-Holland Publ. Comp., Amsterdam (1985).

Grzegorzewski, P.: Testing statistical hypotheses with vague data. Fuzzy Sets and Systems 112(2000), 501–510.

Haas, A.: Zur Theorie der orthogonalen Funktionssysteme. Math. Annalen 69 (1910) 331–371.

Hartigan, J.: Clustering Algorithms. John Wiley & Sons, New York (1975).

Hensel, A.; Spittel, T.: Kraft- und Arbeitsbedarf bildsamer Formgebungsverfahren. Deutscher Verlag für Grundstoffindustrie, Leipzig (1978).

Huber, P. J.: Robust Statistics. John Wiley & Sons, New York (1981).

Huber, P.: Projection pursuits. The Annals of Statistics 13 (1985) 435–525.

Hughes, T. J. R.: The Finite Element Method – Linear Static and Dynamic Finite Element Analysis, Prentice-Hall, Englewood Cliffs, New York 1987.

Jambu, M.: Exploratory and Multivariate Data Analysis. Academic Press, Boston (1991).

Jardin, N.; Sibson, R.: Mathematical Taxonomy. John Wiley & Sons, New York (1971).

Kaufmann, A.; Gupta, M. M.: Introduction to Fuzzy Arithmetic: Theory and Applications. Van Nostrand Reinhold, New York (1985).

Klaua, D.: Über einen zweiten Ansatz zur mehrwertigen Mengenlehre. Monatsber. Deut. Akad. Wiss. Berlin 8 (1966) 161–177.

Klaua, D.: Grundbegriffe einer mehrwertigen Mengenlehre. Monatsber. Deut. Akad. Wiss. Berlin 8 (1966a) 781–802.

Kneschke, A.: Werkstofffluß im Walzspalt beim ebenen Walzen. Neue Hütte 12 (1967) 555–559

Kneschke, A.; Bandemer, H.: Eindimensionale Theorie des Walzvorgangs. In: Zur Mechanik des ebenen Walzens. Freiberger Forschungsheft B 94, Deutscher Verlag für Grundstoffindustrie, Leipzig (1964) 9–75.

Kolmogorov, A. N.: Grundbegriffe der Wahrscheinlichkeitsrechnung. Springer, Berlin (1933).

Kosko, B.: Neural Networks and Fuzzy Systems: A Dynamical Systems Approach to Machine Intelligence. Prentice-Hill, London (1992).

Kruse, R.; Meyer, K. D.: Statistics with Vague Data. Reidel, Dordrecht (1987).

Lawley, D. N.; Maxwell, A. E.: Factor Analysis as a Statistical Method. Butterworth, London (1971).

Lehmann, E. L.: Testing Statistical Hypotheses. 2nd Ed., Wiley & Sons, New York (1986).

Leibniz, G. W.: Brief an Bernoulli vom 3. 12. 1703. In: Mathematische Schriften (Gerhardt, Hgb.), Band III/1, Halle 1855.

Li, Sh.; Ogura, Y.; Kreinovich, V.: Limit Theorems and Application of Set-Valued and Fuzzy Set-Valued Random Variables. Kluwer Academic Publishers, Dordrecht-Boston-London (2002).

Mardia, K. V.; Kent, J. T.; Bibby, J. M.: Multivariate Analysis. Academic Press, London (1979).

Maritz, J. S.: Empirical Bayes Methods. Methuen London (1970).

Matheron, G.: Les Variables Régionalisées et Leur Estimation. Masson et Cie, Éditeurs, Paris (1965).

Matheron, G.: Le Krigeage Universel. Cahiers du Centre de Morphologie Mathematic 1, Fontainbleau (1969).

Matheron, G.: Random Sets and Integral Geometry. John Wiley & Sons, New York (1975).

Minsky, M.; Papert, S.: Perceptron: An Introduction to Computational Geometry. MIT Press, Cambridge, Massachusetts (1969).

von Mises, R.: Grundlagen der Wahrscheinlichkeitsrechnung. Math. Zeitschr. 5 (1919) 52–99.

Moore, R. E.: Methods and Applications of Interval Analysis. SIAM, Philadelphia (1979).

Nagel, M.; Wernecke, K.-D.; Fleischer, W.: Computergestützte Datenanalyse. Verlag Technik, Berlin (1994).

Nahmias, S.: Fuzzy variables in a random environment. In: Advances in Fuzzy Set Theory and Applications (Gupta, M. M.; Ragade, R. K.; Yager, R. R., eds.), North-Holland Publ. Comp., Amsterdam (1979) 165–180.

Nuding, E.: Intervallrechnung und Wirklichkeit. In: Interval Mathematics (Nickel, K., ed.) Lecture Notes in Comp. Science 29, Springer, Berlin (1975).

Ogawa, J.: Statistical Theory of the Analysis of Experimental Design. Marcel Dekker, Inc. New York, (1974).

Otto, M; Bandemer, H.: Calibration with imprecise signals and concentrations based on fuzzy theory. Chemometrics and Intelligent Laboratory Systems 1 (1986) 71–78.

Otto, M.; Bandemer, H.: A fuzzy approach to predicting chemical data from incomplete, uncertain and verbal compound features. In: Physical Property Prediction in Organic Chemistry (Jochum, C.; Hicks, M. G.; Sunkel, J. Eds.), Springer-Verlag, Berlin (1988a) 171–189.

Otto, M.; Bandemer, H.: Fuzzy inference structures for spectral library retrieval systems. Proc. Intern. Workshop on Fuzzy Systems Applications, Iizuka, Fukuoka (1988b).

Polasek, W.: Explorative Datenanalyse. 2. Auflage, Springer, Berlin (1994).

Pilz, J.: Bayesian Estimating and Experimental Design in Linear Regression. John Wiley & Sons, Chichester (1991).

Pilz, J.; Pluch, P.; Spöck, G.: Bayesian kriging with lognormal data and uncertain variogram parameters. in: R. Froideveaux and P. Renard (Eds): geoEnv 2004. Kluwer Academic Publishers Dordrecht 2005.

Puri, M. L.; Ralescu, D.: A possibility measure is not a fuzzy measure. Fuzzy Sets and Systems 7 (1982), 311–313.

Puri, M. L.; Ralescu, D.: Fuzzy random variables. J. Math. Anal. Appl. 114 (1986) 409–422.

Robert, Ch. P.: The Bayesian Choice. Springer, Berlin (2001).

Rojas, R.: Theorie der Neuronalen Netze. Eine systematische Einführung. Springer, Berlin (1993).

Rommelfanger, H.: Fuzzy Decision Support-Systeme. Entscheiden bei Unschärfe. 2. Auflage, Springer, Heidelberg (1994).

Rumelhart, D. E. (ed.): Parallel Distributed Processing. Exploration in the Microstructure of Cognition. Vol. 1: Foundations. The MIT Press Cambridge, Massachusetts (1986).

Ruspini, E.: Numerical methods of fuzzy clustering. Inf. Sci. 6 (1972) 273–284.

Savage, L. J.: The Foundation of Statistics. 2nd Edition. Dover Publications, New York (1972).

Schmerling, S.; Bandemer, H.: Methods to estimate parameters in explicit functional relationships. In: Problems of Evaluation of Functional Relationships from Random-Noise or Fuzzy Data, (Bandemer, H. Ed.), Deutscher Verlag für Grundstoffindustrie, Leipzig (1985).

Schneider, I. (Ed.): Die Entwicklung der Wahrscheinlichkeitstheorie von den Anfängen bis 1933. Akademie-Verlag, Berlin (1989).

Serra, J.: Image Analysis and Mathematical Morphology; 1. Academic Press, New York (1982).

Shafer, G.: A Mathematical Theory of Evidence. Princeton Univ. Press, Princeton (1976).

Smets, Ph.: The degree of belief in a fuzzy event. Information Sci. 25 (1981) 1–19.

Sneath, P. H. A.: The application of computers in taxonomy. J. General Microbiology 17 (1957) 201–226.

Sorenson, T.: A method of establishing groups of equal amplitude in plant sociology on similarity of species content and its application to analysis of the vegetation on Danish common. Biol. Skr. 5 (1968) 1–34.

Stoyan, D.; Kendall, W. S.; Mecke, J.: Stochastic Geometry and Its Applications. Wiley & Sons, Chichester (1995).

Sugeno, M.: Theory of Fuzzy Integral and Its Applications. Ph. D. Thesis, Tokyo Inst. of Technology, Tokyo (1974).

Sugeno, M.: Fuzzy measures and fuzzy integrals: a survey. In: Fuzzy Automata and Decision Processes (Gupta, M. M.; Saridis, G. N.; Gaines, B. N., eds.), North-Holland Publ. Comp., Amsterdam (1977) 89–102.

Tanaka, H.; Uejima, S.; Asai, K.: Linear regression analysis with fuzzy model. IEEE Trans. Systems Man Cybernet. 12 (1982) 903–907.

Tanaka, H.; Watada, J.: Possibilistic linear systems and their application to the linear regression model. Fuzzy Sets and Systems 27 (1988) 275–289.

Toutenburg, H.: Prior Information in Linear Models. John Wiley & Sons, Chichester (1982).

Uhlmann, W.: Kostenoptimale Prüfpläne. Tabellen, Praxis und Theorie eines Verfahrens der statistischen Qualitätskontrolle. 2. Auflage, Physica-Verlag, Würzburg (1970).

Viertl, R.: Statistical Methods for Non-Precise Data. CRC Press, Boca Raton, Florida, (1996).

Viertl, R.; Hareter, D.: Generalized Bayes theorem for non-precise a-priori distributions. Metrika 59 (2004), 263–273.

Walnut, D.: An Introduction to Wavelet Analysis, Birkhäuser, Basel 2002.

Wang, P.-Z.; Sanchez, E.: Treating a fuzzy subset as a projectable random subset. In: Fuzzy Information and Decision Processes (Gupta, M. M.; Sanchez, E., eds.), North-Holland Publ. Comp., Amsterdam (1982) 213–219.

Wishart, D.: Mode analysis: A generalization of nearest neighbour which reduces chaining effects. In: Numerical Taxonomy (Cole, A. J. ed.) Acad. Press, New York (1969), 282–319.

Wold, H.: System analysis by partial least squares. In: Measuring the Unmeasurable, (Nijkamp, P.; Leitner, H.; Wrigley, N. Eds.), Martinus Nijhoff Publ., Dordrecht (1985).

Yager, R. R.: A representation of the probability of a fuzzy subset. Fuzzy Sets and Systems 13 (1984), 273–283.

Zadeh, L. A.: Fuzzy sets. Information and Control 8 (1965) 338–353.

Zadeh, L. A.: Probability measures of fuzzy events. J. Math. Anal. Appl. 23 (1968), 421–427.

Zadeh, L. A.: The concept of a linguistic variable and its application to approximate reasoning I–III. Information Sci 8, 199–250, 301–357; 9, 43–80 (1975).

Zadeh, L. A.: Fuzzy sets as a basis for a theory of possibility. Fuzzy Sets and Systems 1 (1978) 3–28.

Zimmermann, H.-J.: Fuzzy Set Theory and Its Applications. Second Edition, Kluwer-Nijhoff, Dordrecht (1991).

Index

BAYESIAN inference   152
BAYES' estimator
   linear   152
BAYES' theorem   68
   fuzzy analogue of   171
DEMPSTER rule   103
FOURIER series   36
SHANNON entropy   139

a-posteriori probability   68
a-prior probability   68
a-priori distributions   89
accuracy of a confidence region
      estimation   84
accuracy of confidence interval   84
activating function   124
addition of two fuzzy numbers   57
addition of variances
   rule for the   75
addition rule for probabilities of
      mutually disjoint events   66
adequacy of the model   2
aim of an investigation   1, 18
algebraic product   54
algebraic sum   54
algorithm
   genetic   38
alpha-cut   45
analysis
   data   13
approximating polynomials
   structure of the system of   24
approximating setup   33

approximation   21
   global   18, 31
   local   18
   uniform   29
approximation in the quadratic mean
      26
approximation principle   19, 29, 169
assumptions on the data genesis   141
atoms   65

basic probability assignment   102
Bayes' risk   91
belts
   fuzzy transfer principle for   169
betting behaviour and probability   67
binomial distribution   70
body of evidence   102

calibration
   fuzzy   171
Cartesian product
   fuzzy   54
categorial variable   12
certain event   65
certainty of a confidence region
      estimation   84
chance   64
   effect of   64
checking data quality   79, 80
choice
   setup   32
choice of the model   2
choice of the scale   156
classical definition of probability   65

classification   114, 121, 138
cluster   15, 113
  fuzzy   129
cluster analysis   114
cluster validation   114
combination rules
  linguistic   51
comparability of single events   64
compensatory operation   54
complement of a fuzzy set   53
complement of an event   65
components
  minimum-related   105
concrete sample   78
conditional probability   66
conditions
  experimental   64
confidence
  relative level of   102
confidence estimation   151
confidence interval   84
  accuracy of   84
confidence region estimation   83, 84
  accuracy of   84
  certainty of   84
  indetermination relation of   85
conflict
  total   103
conflict between probability assignments
      103
conjugate families of distributions   94
consistency of an estimator   83
contingency table   16
continuous random variable   71
contour
  fuzzy   166
controler
  fuzzy   61
convergence
  speed of   5
convergence in distribution functions
      77
convergence of series of random
      variables   76
core of a fuzzy set   48
covariance function   157
credibility
  degree of   104
credibility of data   14

critical domain   86

data
  manipulation of   14, 80
  pseudo-exact   15
  reliability of   14
data analysis   13
data genesis
  assumptions on the   141
data matrix   114
data quality   79
  checking   79, 80
datum   11
decision on what is the effect of chance
      64
decision theory   68, 91
degree of credibility   104
degree of plausibility   105
degree of possibility   99
delta rule   125
deMorgan laws for fuzzy sets   53
dendrogram   119
density   71
  probability   71
dependence
  type of   19
design of experiments
  orthogonal   25
difference of fuzzy numbers   55
discrete distribution   69
discrete random variable   69
discriminablility of a feature   139
discriminance analysis   121
discriminance procedure   114
discrimination
  model   149
discrimination rule   121
dissimilarity coefficient   115
distance matrix   115
distribution   69
  discrete   69
  Poisson   70
  possibility   99
  probability   69
  binomial   70
distribution family   81
distribution function   73
distribution parameters   70
distribution type   81

distributions
  a-priori   89
division of fuzzy numbers   58
domain
  fuzzy   49

effect of chance   64
  decision on what is the   64
effectivity of an unbiased estimator   83
elementary event   65
empirical distribution function   81
empirical regression   30
empirical variogram   157
equality relation   59
equivalence relation   61
error
  model   149
  numerical   146
  observational   142
  random   142
  relative   147
error of first kind   86
error of second kind   86
error probability   87
error propagation   39
error propagation law   40, 75, 76
essentiality of data   14
estimation
  confidence   151
  confidence region   83, 84
  linear   144
  maximum possibility   106
  unbiased   144
estimation problem
  statistical   144
estimation procedure   82
estimation theory
  main assumption of   82
estimator   82
  consistency of an   83
  effectivity of an unbiased   83
  unbiasedness of an   83
evaluation of a functional relationship
  141
event
  certain   65
  complement of an   65
  elementary   65
  impossible   65

  random   65
events
  comparability of single   64
  intersection of   65
  union of   65
events in masses   63
evidence
  body of   102
evidence weight   103
expected value   74
expected value of the membership
  function   108
experiment
  random   64
  result of an   65
experimental conditions   64
experimental design
  statistical   146
explanatory variables   142
exponential distribution   73
extension principle   55

feature discrimination   114
feature selection   114
feature values
  numerical coding   114
features   114
field
  isotropy of a   157
  random   155
finite elements
  method of   32
finite elements
  main formula of the method of   35
  method of   21, 34
focal set   102
frequency analysis   14
frequentistic interpretation of probability   67
function
  membership   45
  reference   56
function as datum   26
functional relationship   17, 32, 39, 141
  evaluation of a   141
fuzzily expressed fuzzy similarity   135
fuzzily expressed similarity   138
fuzziness measure   172
fuzzy analogue of BAYES'theorem   171

fuzzy calibration   171
fuzzy Cartesian product   54
fuzzy cluster   129
fuzzy contour   166
fuzzy controler   61
fuzzy domain   49
fuzzy interpolation   170
fuzzy interval   49, 55
fuzzy measure   97, 101, 102
fuzzy measures   107
  specification of   101
fuzzy n-partition   130
fuzzy neighbourhood of a fuzzy set
      136
fuzzy neural networks   128
fuzzy number   48, 55, 88
  negative of   55
  reciprocal of a   56
fuzzy numbers
  addition of two   57
  difference of   55
  division of   58
  product of two   58
  sum of   55
fuzzy point   49
fuzzy region   166
fuzzy relation   59
fuzzy set   43, 45
  complement of a   53
  core of a   48
  inducing   99
  support of a   48
fuzzy sets
  intersection of two   53
  product of   55
  quotient of   56
  random   111
  theory of   43
  union of two   53
fuzzy similarity   135, 140
fuzzy similarity of all objects   139
fuzzy similarity relation   133
fuzzy transfer principle for belts   169
fuzzy variable   50
fuzzy vector   49

genetic algorithm   38
geometrical probability   66
glancing intersection   24

global approximation   18, 31
global interpolation   18
golden rule of problem solving   8
grades of a property   12
grey tone picture as datum   13
grey tones as membership values   49

hidden units   124
histogram   81
HPD region   91
hypothesis   85, 86

ignorance
  partial   102
  total   102
impossible event   65
inclusion of a fuzzy set in another one
      53
inclusion of an event in another one
      65
independence
  pairwise   68
independence of events   69
independence of the components of a
      random vector   79
independently obtained realizations
      81
indetermination relation of confidence
      region estimation   85
inducing fuzzy set   99
inference
  possibilistic   105
information processing   8
information situation   18
information theory   68
instability
  numerical   24
interpolation   19
  fuzzy   170
  global   18
  linear   19
  local   18
interpolation of higher order   20
intersection
  glancing   24
intersection of events   65
intersection of two fuzzy sets   53
interval
  fuzzy   49, 55

resulting   41
interval arithmetics   41, 56
interval mathematics   41, 42
intrinsic hypothesis   157
isotropy of a field   157

knowledge base   138
Kriging-technique   158

law
  error propagation   40
law of large numbers   76
learning theory   68
likelihood function   121
limit theorem   77
linear estimation   144
linear interpolation   19
linear setup   143
linear BAYES' estimator   152
linguistic combination rules   51
linguistic modifier   51
linguistic variable   50
local approximation   18
local interpolation   18
local monoticity   47
local smoothing   30
local validity of a model   156

main assumption of estimation theory
  82
main formula of the method of finite
  elements   35
manipulation of data   14, 80
mathematical sample   79
mathematical statistics   78
mathematical structures   15
maximum possibility estimation   106
mean membership degree   109
measure   97
  fuzziness   172
  fuzzy   97, 102, 107
  necessity   100
  possibility   99
membership function   45
  specification of a   47
  triangular   58
membership values
  grey tones as   49
method of finite elements   21, 32, 34

method of least squares   145
method of principal components   150
minimum-related components   105
modal analysis   120
modal trace   161
model
  probabilistic   77
model adequacy   2
model choice   2
model discrimination   149
model error   149
model harmony   7
modifier
  linguistic   51
monoticity
  local   47
multi-dimensional random variable   76
multi-phase regression   33
multiplication of a fuzzy number by a
  crisp positive number   57
mutually disjoint events
  addition rule for probabilities of   66

necessity measure   100
negative of a fuzzy number   55
neighbourhood of a feature   138
neighbourhood of a fuzzy set   136
neighbourhood of an object   138
networks
  fuzzy neural   128
neural networks   68, 122
  fuzzy   128
neurons   122
nominal variable   12
non-located element   107
normal distribution   71
normalized measure
  probability as a   66
not localized element   98
nugget-effect   158
number
  fuzzy   48, 55
numerical coding of feature values   114
numerical error   146
numerical instability   24
numerical stability   5

observational error   142
one-dimensional random variable   69

operation
  compensatory   54
opinion of an expert as datum   13
ordinal variable   12
orthogonal design of experiments   25
orthogonal polynomials   24, 27
orthogonal system of functions   27
outcome of a trial   64
outlier   5, 14, 79, 106, 147

P-value for testing hypothesis   87
pairwise independence   68
parameter estimation   82
partial ignorance   102
partial least squares technique   160
partition matrix   129
pattern cognition   15
pattern in data   15
plausibility
  degree of   105
point
  fuzzy   49
Poisson distribution   70
polynomials
  orthogonal   24, 27
possibilistic inference   105
possibilistic variable   105
possibility
  degree of   99
possibility degree   99
possibility distribution   99
possibility measure   99
preciseness of data   6
previous knowledge   89
principal components
  method of   150
principle
  approximation   19
principle of the smallest sum of the
    absolute values   29
probabilistic model   77
probability
  a-posteriori   68
  a-priori   68
  betting behaviour and   67
  classical definition of   65
  conditional   66
  frequentistic interpretation of   67
  geometrical   66

  specification of subjective   67
  subjective interpretation of   67
probability as a normalized measure
    66
probability as the limit of relative
    frequency   66
probability assignment
  basic   102
probability assignments
  conflict between   103
probability density   71
probability distribution   69
probability theory   66
problem
  specification   46
  structure   46
problem solving
  golden rule of   8
procedure impreciseness   87
process as variable   12
process control   86
product of fuzzy sets   55
product of two fuzzy numbers   58
projection pursuit   17, 160
propagation
  error   39
property
  grades of a   12
pseudo-exact data   15, 87

quadratic mean
  approximation in the   26
qualification
  rules for   51
quality control   85
quality requirements   1
quantification
  rules for   51
quantitative variable   12
quotient of fuzzy sets   56

random error   142
random event   65
random experiment   64
random field   155
random fuzzy sets   111
random interval   84
random variable   78, 142
  continuous   71

discrete   69
  multi-dimensional   76
  one-dimensional   69
  realization of a   78
random variables
  series of   76
randomization   80
randomness   156
realization of a random variable   78
realization of a variable   11
realizations
  independently obtained   81
reciprocal of a fuzzy number   56
reference function   56
region
  fuzzy   166
regionalized variable   12, 155
regression
  empirical   30
  multi-phase   33
regression analysis   144
regression lines   153
regression of second kind   153
regularities of events in masses   63
relation
  equality   59
  equivalence   61
  fuzzy   59
  fuzzy similarity   133
  similarity   60
relationship
  functional   17, 32, 39, 141
relative error   147
relative frequency
  probability as the limit of   66
relative level of confidence   102
reliability of data   14
repeatability of conditions   64
response surface   142
response value   142
response variable   142
result of an experiment   65
resulting interval   41
robust statistics   6
robustness   5, 93
rule for the addition of variances   75
rule of weakest chain-link   2
rules for qualification   51
rules for quantification   51

sample   78
  concrete   78
  mathematical   79
  truncated   80
sample size   79
sample space   78
scale
  choice of   156
search procedures   38
semivariogram   157
sequential procedures   93
series of random variables   76
set
  fuzzy   45
setup   31, 141, 143
  approximating   33
  linear   143
  true   144, 148
setup choice   32
sigmoid function   124
similarity   115
  fuzzy   135, 140
similarity concept   133
similarity degree   115, 137
similarity matrix   115
similarity relation   60, 115
  fuzzy   133
similarity structure of the knowledge
    base   138, 139
similarity threshold   117
smoothing
  local   30
specification of a membership function
    47
specification of fuzzy measures   101
specification of subjective probability
    67
specification problem   46
speed of convergence   5
spline technique   21
stability
  numerical   5
statistical estimation problem   144
statistical experimental design   146
statistical interpretation of the
    probability of a fuzzy set   108
statistics
  robust   6

structur of a system of approximating
    polynomials   24
structure problem   46
structures
  mathematical   15
subjective interpretation of probability
    67
subtraction of fuzzy numbers   58
sum of fuzzy numbers   55
support of a fuzzy set   48
system of approximating polynomials
  structure of a   24
system of functions
  orthogonal   27

test variable   86
testing hypothesis
  P-value for   87
theory of fuzzy sets   43
threshold function   124
total conflict   103
total enumeration   119
total ignorance   102
transfer principle   162
transformation of data   15
trial
  outcome of a   64
triangular membership function   58
true model   82
true setup   144, 148
truncated sample   80
trustworthiness of data   14
type of dependence   19
type of distribution   78

unbiased estimation   144
unbiasedness of an estimator   83

unbiasness   144
uncorrelatedness   143
uniform approximation   29
union of events   65
union of two fuzzy sets   53
universal set   44
universe   44
universe of discourse   11

validity of a model
  local   156
variable
  categorial   12
  fuzzy   50
  linguistic   50
  nominal   12
  ordinal   12
  possibilistic   105
  quantitative   12
  realization of a   11
  regionalized   12, 155
variance   75
variances
  rule for the addition of   75
variedness of the features   139
variogram   157
  empirical   157
vector
  fuzzy   49

wavelets   32, 37
weakest chain-link
  rule of the   2
weight
  evidence   103
window   30

Printed in the United States
63077LVS00002B/272